Library of
Davidson College

Applied Demography for Biologists

Applied Demography for Biologists

with Special Emphasis on Insects

JAMES R. CAREY
University of California, Davis

New York   Oxford
OXFORD UNIVERSITY PRESS
1993

Oxford University Press

Oxford   New York   Toronto
Delhi   Bombay   Calcutta   Madras   Karachi
Kuala Lumpur   Singapore   Hong Kong   Tokyo
Nairobi   Dar es Salaam   Cape Town
Melbourne   Auckland   Madrid

and associated companies in
Berlin   Ibadan

Copyright © 1993 by Oxford University Press, Inc.

Published by Oxford University Press, Inc.,
200 Madison Avenue, New York, New York 10016

Oxford is a registered trademark of Oxford University Press.

All rights reserved. No part of this publication may be reproduced,
stored in a retrieval system, or transmitted, in any form or by any means,
electronic, mechanical, photocopying, recording, or otherwise,
without the prior permission of Oxford University Press.

Library of Congress Cataloging-in-Publication Data
Carey, James R.
Applied demography for biologists with special emphasis on insects/
by James R. Carey.
p.   cm.   Includes bibliographical references and index.
ISBN 0-19-506687-1
1. Population biology.   2. Demography.   I. Title.
QH352.C36   1993   574.5'248'072—dc20   91-45843

9 8 7 6 5 4 3 2 1

Printed in the United States of America
on acid-free paper

To my family

Preface

The idea for this book stemmed from my Ph.D. research on the demography of several species of Acari including life tables, population dynamics, and age structure. While writing my dissertation I discovered that the only books available on demography for biologists were ecology texts that offered simple examples of life table construction and computation of the intrinsic rate of increase. These texts generally did not cover, for example, the meaning of the stable age distribution or a discussion of its importance. Therefore, I decided that biologists needed a stand-alone book on demography that went beyond the standard life table and, among other things, *did* include a detailed treatment of the stable age distribution.

The book itself began as a syllabus for a course I first offered at UC Davis in the mid-1980s entitled "Insect Demography." Most graduate students who enrolled in my course did so largely because their research involved some aspect of demography—birth, death, or population. I have the same objectives for this book as I did for the course from which it was developed— to provide an analytical and conceptual foundation for demographic aspects of biological and ecological research. My emphasis on insects is unavoidable because I am an entomologist. However, demographic principles covered in the book are not restricted to insects any more than techniques presented in books on human demography are restricted to *Homo sapiens*. I believe the focus on a specific group of organisms, in this case insects, provides continuity and also eliminates the use of token examples. That is, I draw heavily from detailed demographic data from my own research or that of my students.

The book has four main themes—birth, death, population, and application—and is organized in six chapters. Chapter 1 is simply an introduction to demography. Here I define terms, provide background concepts, and introduce elementary formulae.

Chapter 2 covers the main techniques of life table analysis, including the current and abridged single decrement tables, the multiple decrement life table, and basic life table properties such as sensitivity analysis and statistics. I included the sections on elementary life table statistics because I felt that it is useful to establish the connection between the life table and conventional statistics. This relationship was not obvious to me when I first studied demography and apparently it was not obvious initially to many demographers and statisticians—the first life table was published in 1662 yet statistical aspects were not formalized until the mid-1960s.

Chapter 3 on reproduction was taken almost exclusively from my own work which was strongly influenced by reading the demographic literature on reproduction. This chapter covers basic concepts of analysis of reproductive heterogeneity and parity-specific life tables. My emphasis here is mostly on organization and analysis of data on reproduction and not on mathematical models, per se.

I devote two chapters to analysis of population. The first of these, Chapter 4, covers the population basics including elementary models and concepts such as rates of increase. At the end of this chapter I include a section on basic properties of population such as population momentum and convergence to the stable age distribution. I omitted the mathematics on these topics because I felt that the concepts were far more important than were the mathematical details. However, I cite several of the key works in each of these sections for those who wish to follow up on the mathematics.

A treatment of more advanced population models is contained in Chapter 5 where I cover four important extensions of stable population theory—two-sex models, stochastic demography, multiregional demography, and demographic theory for social insects. With the exception of stochastic models, all of these models fall into the category demographers refer to as multidimensional demography. At the end of this chapter I include a summary and synthesis of the main properties of stable models—the tendency of models with fixed rates to converge to identical (or similar) states that are independent of initial conditions.

In Chapter 6, the last chapter, I cover a number of interconnections and applications of demographic tools and principles including what I consider one of the most realistic applications of demographic theory—harvesting concepts applied to problems in mass rearing.

I included as an appendix a life table constructed from the mortality experience of over 1.2 million medfly adults. Funded by the National Institute on Aging, the study was designed to examine age-specific mortality in the oldest-old individuals and is the largest life table for any nonhuman species. Therefore I felt that making it available in this book would be particularly useful to biologists.

I am indebted to many people who helped make this a better book. I particularly thank the students who enrolled in my insect demography course at UC Davis over the last six years and whose feedback kept me focused on actual users. I gratefully acknowledge the input and data from many of my graduate students including David Berrigan, Andrew Hamilton, David Krainacker, Pablo Liedo, Rachael Freeman, Marc Tatar, and Pingjun Yang. I thank Wayne Getz and Gregory Payne for their help in the early stages of the book formation. The recent books by Hal Caswell, *Matrix Population Models*, and Shripad Tuljapurkar, *Population Dynamics in Variable Environments*, were particularly helpful for writing several key sections on population.

Appreciation is extended to members of the Department of Demography at UC Berkeley, particularly Eugene Hammel, Kenneth Wachter, and Ronald Lee, for allowing me to participate in their seminars and interact with

PREFACE

their group. I thank Griffith Feeney for providing many useful demographic reprints and offprints and James Vaupel for sharing his insights on mortality analysis and heterogeneity. I am grateful to Shripad Tuljapurkar for his direct input to sections on stochastic demography and on the demography of social insects. I gratefully acknowledge the efforts of William Curtis, who was instrumental and especially encouraging in the early stages of development, and Kirk Jensen at Oxford University Press for his encouragement and advice in the later stages. Appreciation is extended to my colleagues in the Department of Entomology, UC Davis, for their support. I thank Howard Ferris who was the first person to encourage me to develop my course syllabus into a book and Robert Washino who was particularly supportive as Department Chair during the formative stages. Lastly I extend deep appreciation to my wife, Nancy, and my children Bryce, Ian, and Meredith, for their understanding, patience, and encouragement.

Davis, California J. R. C.
February 1992

"The members of any cohort are entitled to participate in only one slice of life—their unique location in the stream of history."
                         Norman Ryder (1965)

Contents

1. **Introduction, 3**

    Formalization, 5
    Demographic Levels and Traits, 5
    The Life Course, 5
    Demographic Rates, 5
    Elementary Characteristics of Populations, 7
    Population Size, 7
    Population Distribution, 7
    Population Structure, 8
    Population Change (in Size), 8
    Population Change (in Space), 9
    Bibliography, 9

2. **Life Tables, 11**

    General Concepts, 11
    The Single Decrement Life Table, 11
    Life Table Radix, 11
    Life Table Functions, 12
    Complete Cohort Life Table, 13
    The Abridged Life Table, 19
    The Multiple Decrement Life Table, 22
    Data and Data Organization, 23
    General Framework and Notation, 24
    Table Construction, 25
    Elimination of Cause—Concept, 26
    Elimination of Cause—Application, 29
    General Concepts, 31
    Selected Properties of Model Life Tables, 33
    Life Table Statistics, 33
    Sensitivity Analysis, 38
    Life Table Entropy, 39
    Bibliography, 41

3. Reproduction, 43

General Background, 44
Per Capita Reproductive Rates, 44
Daily Rates—Basic Notation and Schedules, 44
Lifetime Rates and Mean Reproductive Ages, 50
Previous and Remaining Reproduction at Age x, 56
Reproductive Heterogeneity, 57
Birth Intervals, 57
Reproductive Parity, 64
Generalizations, 67
Clutch, 67
Reproductive Life Table and Parity Progression, 71
Bibliography, 75

4. Population I: Basic Concepts and Models, 77

Background, 77
Population Rates, 77
The Balancing Equation, 78
Doubling Time, 79
Growth with Subpopulations, 79
Sequence of Growth Rates, 80
The Stable Population Model, 81
Derivation, 82
Intrinsic Rate of Increase, 83
Intrinsic Rate of Increase: Analytical Approximations, 85
Net Reproductive Rate, 87
Intrinsic Birth and Death Rates, 87
The Stable Age Distribution, 89
Mean Generation Time, 91
Reproductive Value, 91
Population Projection, 92
Leslie Matrix, 92
Example Iteration, 95
Fundamental Properties of Populations, 99
Age Structure Transience, 99
Independence of Initial Conditions, 100
Fertility and Mortality, 101
Changing Schedules and Unchanging Age Structure, 102
Effect of Age of First Reproduction on Population Growth Rate, 102
Speed of Convergence, 103

CONTENTS

Population Momentum, 104
Bibliography, 104

5. Population II: Extensions of Stable Theory, 106

Two-sex Models, 106
Basic Two-Sex Parameters, 106
Sex Ratio at Age x, 109
Sex Ratio by Stage, 111
Overall Sex Ratio, 112
Stochastic Demography, 113
General Background, 113
Environmental Variation in Vital Rates, 114
Stochastic Rate of Growth, 115
Multiregional Demography, 118
Location Aggregated by Birth Origin, 119
Location Disaggregated by Birth Origin, 120
Age-by-Region Projection Matrix, 123
Demographic Theory of Social Insects: The Honeybee, 124
Concept of a Superorganism, 124
Growth Limits, 126
Within-Colony Dynamics, 128
Swarming and Generation Time, 129
Colony Demography, 131
Application, 132
The Unity of Demographic Population Models, 134
Leslie Matrix Models, 134
Models Structured by Stage or Size, 134
Stationary Population Models, 134
Stochastically Varying Vital Rates, 134
Deterministically Varying Vital Rates, 135
Two-Sex Model, 135
Multiregional Models, 135
Models Structured by Age and Genotype, 135
Hierarchical Population Model (Honeybees), 135
Models of Kinship, 136
Parity Projection Models, 136
Bibliography, 137

6. Demographic Applications, 140

Estimation, 140
Growth Rate, 140

Insect-Days, 142
Stage Duration, 145
Mosquito Survival Analysis Using Parity Rates, 148
Pesticide Effective Kill Rate, 150
Probit Analysis; Life Table Perspectives, 152
Percent Parasitism, 154
Life History Scaling, 155
Curve Fitting, 155
The Linear Equation: Least Squares, 157
Population Curve: The Logistic Equation, 158
Mortality Curve: The Gompertz, 159
Fecundity Curve: The Pearson Type I, 161
Mass Rearing: Basic Harvesting Concepts, 161
Medfly, 162
Spider Mites, 164
Parasitoid Mass Rearing, 168
Harvest Rates, 169
Per Female Production Rates, 170
Host-Parasitoid Coupling, 172
Age and Species Structure, 176
Stage Structure, 177
Bibliography, 180

Appendix 1: A Preliminary Analysis of Mortality in 1.2 Million Medflies, 182
Background, 182
Life Table Functions, 183
Male and Female Age-Specific Mortality, 184
Mortality Rate Doubling Time, 185
Crossovers, 186
Age Groups Responsible for Sex Differences in Life Expectancies, 188
Effect of Proportional Mortality Differences on Life Expectancy, 189
Life Table Entropy, 190
The Force of Mortality and Oldest-Old, 191
Estimation, 192
Bibliography, 195

Appendix 2: Life Table for 598,118 Male Medflies, 196
Appendix 3: Life Table for 605,528 Female Medflies, 200

Applied Demography for Biologists

1
Introduction

All biological populations obey the immutable rule that, migration withstanding, the difference between their birth and death rates determines the direction and size of their growth rate. This relationship provides the foundation for demography—the study of populations and the processes that shape them (Pressat, 1985).

Demography began as the study of *human* populations and literally means "description of the people." The word is derived from the Greek root *demos*, meaning "the people," and was coined by a Belgian, Achille Guillard, in 1855 as *demographie* (*Elements of Human Statistics or Comparative Demography*). He defined demography as the natural and social history of the human species or the mathematical knowledge of populations, of their general changes, and of their physical, civil, intellectual, and moral condition (Shryock et al., 1976).

Although demography is a distinct discipline in its own right, only a few academic departments or graduate groups exist worldwide that are devoted exclusively to demography. Aside from census bureau personnel, the vast majority of researchers or practitioners who use demographic methods are part of more broadly defined disciplines such as sociology, psychology, business, and medicine for human populations or ecology, fisheries, wildlife, forestry, and entomology for plant and animal populations. In principle both demography as a field and population as an entity can be defined in the abstract and thus encompass the social as well as the biological sciences. However, in practice this is not the case. Every field creates definitions that fits its needs. Social scientists tend to refine versions of Guillard's original definition of demography. Hence populations that represent their center of relevance are almost always human and are typically defined in geopolitical terms.

Biologists typically define *population* in Mendelian terms such as "a group of interbreeding organisms belonging to the same species and occupying a clearly delimited space at the same time" (Wilson, 1975) or "...a cluster of individuals with a high probability of mating with each other compared to their probability of mating with a member of some other population" (Pianka, 1978). In many applied areas of biology the term *population* often refers to an entity about which statistical information is desired. For example, sampling leaves for insect infestation levels is viewed in statistics as sampling from a population (i.e., the totality of elements about which information is desired). Samples from this target "population" will generate a mean and variance and in turn enable one to test statistical hypotheses. Population in this sense may be relevant to demographers though it may have little to do with an interbreeding group of individuals in the biological sense. Thus it is often necessary to move back and forth between the concept of a population as a material aggregate and population as a biologically reproducing entity. *Population* is defined here in general terms simply as "a group of individuals coexisting at a given moment" (Pressat, 1985).

Classical demography is concerned basically with four aspects of populations (Shryock et al., 1976): (1) *size*—the number of units (organisms) in the population; (2) *distribution*—the arrangement of the population in space at a given time; (3) *structure*—the distribution of the population among its sex and age groupings; and (4) *change*—the growth or decline of the total population or one of its structural units. The first three—size, distribution, and structure—are referred to as population statics, while the last—change—is referred to as the population dynamics. Hauser and Duncan (1959) regard the field of demography as consisting of two parts: i) *formal demography*—a narrow scope confined to the study of components of population variation and change (i.e., births, deaths, and migration); and ii) *population studies*—a broader scope concerned with population variables as well as other variables. In a biological context these other variables may include genetics, behavior, and other aspects of an organism's biology. The methodology of demographic studies includes data collection, demographic analysis, and data interpretation.

In biology it is often difficult to determine where one field leaves off and the other begins. But many apply demographic methods or use the concepts in some way. *Ecology* is concerned with the interrelationship of organisms and their surroundings; *population biology* is often used interchangeably with ecology but usually implies an emphasis on evolutionary relations; *population ecology* is distinct from *community ecology* in that the former is usually concerned with the interactions of a few species, while the latter with the interactions of many; *population dynamics* implies the study of the mechanisms and consequences of population change (usually numbers); *population genetics* is less concerned with numbers of individuals in a population and more concerned with gene frequencies and their rate of change, and *applied ecology* is a rubric for areas such as forestry, fisheries, and pest management.

Demographers conceive the population as the singular object for scientific analysis and research. However, as Pressat (1970) notes, population is everywhere and nowhere in the sense that many aspects of demography can be studied simply as component parts of the disciplines considered. He states, "But to bring together all the theories on population considered as a collection of individuals subject to process of evolution, has the advantage of throwing into relief the many interactions which activate a population and the varied characteristics of that population."

FORMALIZATION

Demographic Levels and Traits

The basic unit and starting point for demographic analysis is the *individual*, which, according to Willekens (1986), is defined simply as "a single organism that is a carrier of demographic attributes." The individual is a natural unit and need not be contrived like "power" or "community." The basic attributes of individuals include a development rate, an age-specific level of reproduction, and a time of death.

The next demographic level for which traits are considered is that of the *cohort*, defined as "a group of same-aged individuals" or, more generally, "a group who experience the same significant event in a particular time period, and who can thus be identified as a group for subsequent analysis" (Pressat, 1985). Cohort attributes are to be distinguished from individual traits in that they possess a mean and variance and therefore are statistical. These traits are often expressed in the form of an age schedule.

The age schedule of events in cohorts determines their *population traits*—the third demographic level. These traits result from the interplay of cohort attributes that are, in turn, set by the individuals within the cohort.

The Life Course

A universal constant for all life is chronological age, which is the exact difference between the time on which the calculation is made and the time of the individual's birth. This difference is typically termed *exact age* and is to be contrasted with *age class*, which groups exact ages into periods. The passage from one stage to another is formally termed an *event*. The sequence of events and the duration of intervening stages throughout the life of the organism is termed its *life course*.

It is clear that age is an important dimension of life and grouping individuals into age classes or at least distinguishing between young and old is useful. Therefore I denote the age of egg hatch as η (eta), age of pupation as π (pi), age of eclosion (first day of adulthood) as ε (epsilon), age of first reproduction as α (alpha), age of last reproduction as β (beta), and oldest possible age as ω (omega), Several key intervals can then be defined using this notion and are summarized in Table 1-1.

This scheme is depicted graphically in Figure 1-1. The notation for adult traits, particularly α, β, and ω, is a convention in demography, while the notation for the preadult is not. In general, age is the characteristic, central variable in almost all demographic analysis and serves as a surrogate for more fundamental measures (e.g., physiological state) and duration of exposure to risk.

Demographic Rates

Demographic rates can be grouped into five categories according either to the kind of population counted in the denominator or to the kind of events

Table 1-1. Summary of Key Intervals in an Insect Life Course

Interval	Notation
Preadult	
Egg incubation	$0-\eta$
Nymphal or larval period	$\eta-\pi$
Pupal period	$\pi-\varepsilon$
Adult	
Preovipositional period	$\varepsilon-\alpha$
Reproductive period	$\alpha-\beta$
Postreproductive period	$\beta-\omega$
Adult life span	$\varepsilon-\omega$
Total life span	$0-\omega$

counted in the numerator (Ross, 1982). These include—

1. *Crude rates.* This category uses the total population as the denominator. Thus we may consider crude birth rate (number of births per number in the population), crude death rate (number of deaths per number in population), and crude rate of natural increase (difference between crude birth and death rates). The results are "crude" in that they consider all individuals rather than by age or sex groupings.
2. *Age-specific rates.* These are the same as crude rates except with age restrictions for both numerator and denominator. For example, age-specific fecundity for an insect 20 days old counts only offspring produced by females in that age group and counts only females of that age in the denominator. Thus a *schedule* of rates is created.
3. *Restricted rates.* These rates apply to any special sub-group. For example, in human demography the number of births to married women (rather than to all women) is termed "marital age-specific fertility rate." In insects a restricted rate of this sort could be fecundity rates of females that produce at least one offspring. This would therefore exclude all steriles.
4. *Rates by topic.* These rates apply to each specialized topic in demo-

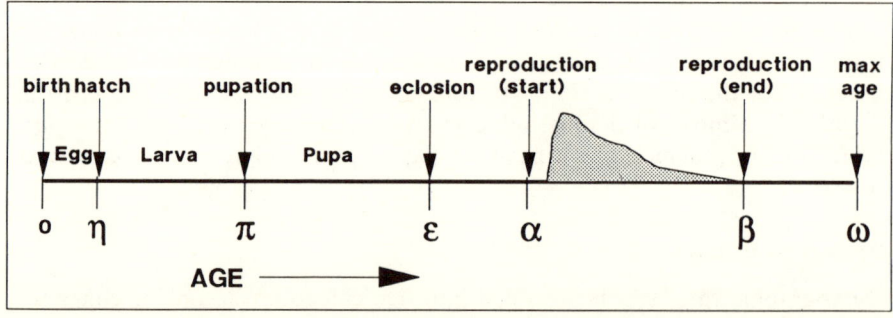

Figure 1-1. Generic diagram of an insect life course.

INTRODUCTION

graphy. For example, reproductive rates may include total fertility rate, gross reproductive rate, or net reproductive rate.

5. *Intrinsic rates.* The rates that prevail in a stable population are referred to as intrinsic rates in that they cannot reflect any accidental or transient short-term feature of the age distribution. Thus they are considered "intrinsic" or "true" rates.

ELEMENTARY CHARACTERISTICS OF POPULATIONS

Information on populations is obtained either through a census or through a survey, the distinction which is far from clear-cut. A census is typically thought of as a complete canvass of an area; the intent is to enumerate every individual in the population by direct counting and further, to cross-classify by age (stage), sex, and so forth. The intent of a survey is to estimate population characteristics on a sample basis.

Population Size

In concept, the notion of population size is extremely simple since it means the total number of individuals in the population. However, human demographers make a distinction between *de facto* enumeration, which records where each individual is at the time of the census (i.e., includes military or migrant workers), and *de jure* enumeration, which records usual residence (i.e., only records military personnel at place of residence).

Population Distribution

There are basically three broad spatial measures that characterize a particular distribution. These are—

1. *Number by spatial subdivision.* Statistics in this case can be given as i) percentage of the total by subdivision, or ii) a rank order from the subdivision with the highest count to the spatial unit with the lowest. Depending upon which method is used, comparisons of two census times will reveal the change in percentage or the change in rank by spatial location.
2. *Measures of central location.* The center of a population or the mean point of the population distributed over an area is defined as the center of population gravity or of population mass. The formula for the coordinates of the population center is given by

$$x = \sum_{i=1}^{k} p_i x_i / \sum_{i=1}^{k} p_i \tag{1-1a}$$

$$y = \sum_{i=1}^{k} p_i y_i / \sum_{i=1}^{k} p_i \tag{1-1b}$$

where p_i is the number in the population at point i and x_i and y_i are its horizontal and vertical coordinates, respectively. Population center can also be defined in three-dimensional space (e.g., vertical distribution on plants) by adding the z-coordinate and computing z.

3. *Measures of concentration and spacing.* The simplest measure of the degree of dispersion of a population in the xy-plane is known as the *standard distance*. This measure bears the same kind of relationship to the center of the population that the standard deviation of any frequency distribution bears to the arithmetic mean. If x and y are the coordinates of the population center, then the standard distance, D, is given by

$$D = \left[\frac{\sum_{i=1}^{k} \{f_i(x_i - x)^2\}}{n} + \frac{\sum_{i=1}^{k} \{f_i(y_i - y)^2\}}{n} \right]^{1/2} \quad (1\text{-}2)$$

where f_i is the number of organisms in a particular area and $n = \sum_{i=1}^{k} f_i$.

Population Structure

The structure of a population is the relative frequency of any enumerable or measurable characteristic, quality, trait, attribute, or variable observed for individuals (Ryder, 1964). These items could include age, sex, genetic constitution, weight, length, shape, color, biotype, birth origin, and spatial distribution. Only age and sex will be covered here since they are the most common traits by which individuals in populations are decomposed.

Sex ratio (SR) is the principal measure of sex composition and is usually defined as the number of males per female or

$$SR = n_m/n_f \quad (1\text{-}3)$$

where n_m and n_f represent the number of males and the number of females, respectively. The proportion of males (PM) in a population is given by

$$PM = n_m/n_T \quad (1\text{-}4)$$

where $n_T = n_m + n_f$. This measure expresses males as a fraction of the total and not as a ratio in the conventional sense. Additional measures of sex composition in ecology include the *primary sex ratio* (sex ratio at conception or birth) and the *secondary sex ratio* (sex ratio at adulthood or at the end of parental care).

The simplest kind of analysis of age or stage data is the frequency distribution of the total population by age or

$$f_x = n_x/N \quad (1\text{-}5)$$

where f_x is the frequency of individual aged x, n_x is the number in the population at age x, and N is the total number in the population. An *age*

Population Change (in Size)

If a population numbers N_t and N_{t+1} at times t and t + 1, respectively, then the *amount* of change equals

$$N_t - N_{t+1} = \text{amount of change} \tag{1-6a}$$

which is simply the difference in the population number at the two time periods. However, the *rate* of change is given by

$$N_{t+1}/N_t = \text{total rate of change} \tag{1-6b}$$

which gives the *factor* by which the population changed over one time period relative to the number at time *t* and

$$(N_{t+1} - N_t)/N_t = \text{fractional rate of change} \tag{1-6c}$$

which gives the *fraction* by which the population changes over one time period relative to the number at time *t*.

Population Change (in Space)

Population change occurs when *migrants* move from one area to another (Shryock and Siegel 1976). Every move is an out-migration with respect to the area of ogirin and an in-migration with respect to area of destination. The balance between in-migration and out-migration is termed *net migration*. The sum total of migrants moving in one direction or the other is termed *gross in-migration* or *gross out-migration*. The sum total of both in- and out-migrations is termed *turnover*. A group of migrants having a common origin and destination is termed a *migration stream*. The difference between a stream and its counterstream is the *net stream* or *net interchange* between two areas. The sum of the stream and the counterstream is called the *gross interchange* between the two areas.

Various rates can be expressed arithmetically as

$$\text{mobility rate} = M/P \tag{1-7}$$

where M denotes the number of movers and P denotes the population at risk of moving. Other formulae for movement include

$$\text{in-migration rate} = M_i = I/P \tag{1-8a}$$
$$\text{out-migration rate} = M_0 = O/P \tag{1-8b}$$
$$\text{net migration rate} = (I - O)/P \tag{1-8c}$$

where I and O denote the number of in-migrants and out-migrants, respectively.

BIBLIOGRAPHY

Hauser, P. M., and O. D. Duncan (1959) The nature of demography. In: *The Study of Population*. P. M. Hauser and O. D. Duncan (eds.), pp. 29–44. The University of Chicago Press, Chicago.
Pianka, E. R. (1978) *Evolutionary Ecology*. 2nd ed. Harper & Row, Publishers, New York.
Pressat, R. (1970) *Population*. C. A. Watts & Co., Ltd., London.
Pressat, R. (1985) *The Dictionary of Demography*. Bell and Bain, Ltd., Glasgow.
Ross, J. A. (1982) *International Encyclopedia of Population*. Vol. 2. The Free Press, New York.
Ryder, N B. (1964) Notes on the concept of a population. *Amer. J. Sociol.* 5:447–463.
Ryder, N. B. (1965) The cohort as a concept in the study of social change. *Amer. Soc. Review* 30:843–861.
Shryock, H. S., J. S. Siegel, and E. G. Stockwell (1976) *The Methods and Materials of Demography*. Academic Press, New York.
Willekens, F. (1986) Synthetic biographies: A method for life course analysis. Paper presented at the Population Association of American meetings, San Francisco.
Wilson, E. O. (1975) *Sociobiology: The New Synthesis*. Belknap Press of Harvard University Press, Cambridge.

2
Life Tables

GENERAL CONCEPTS

A life table is a detailed description of the mortality of a population giving the probability of dying and various other statistics at each age (Pressat, 1985). There are two general forms of the life table. The first is the *cohort life table*, which provides a longitudinal perspective in that it includes the mortality experience of a particular cohort from the moment of birth through consecutive ages until no individuals remain in the original cohort. The second basic form is the *current life table*, which is cross-sectional. This table assumes a hypothetical cohort subject throughout its lifetime to the age-specific mortality rates prevailing for the actual population over a specified period. These are often referred to in ecology as time-specific life tables and are used to construct a *synthetic* cohort.

Both cohort and current life tables may be either *complete* or *abridged*. In a complete life table the functions are computed for each day of life. An abridged life table deals with age intervals greater than one day, such as over a complete stage (e.g., larval period), where precise determination of daily survival is difficult. The distinction between complete and abridged has to do with the length of the age interval considered. Both forms of the life table may be either *single decrement* or *multiple decrement*. The first of these lumps all forms of death into one and the second disaggregates death by cause.

My objective in this chapter is to introduce the basic concepts, notation, and methods of life table analysis, including complete and abridged tables; special properties of the life table, such as sensitivity analysis; and the multiple decrement life table. More advanced treatments of life tables include the works by Brass (1971), Keyfitz and Frauenthal (1975), Schoen (1975), Batten (1978), Elandt-Johnson and Johnson (1980), Chiang (1984), Manton and Stallard (1984), Hakkert (1987), and Pollard (1988).

THE SINGLE DECREMENT LIFE TABLE

Life Table Radix

A radix in mathematics is a number that is arbitrarily made the fundamental number of a system of numbers. In the life table the radix is the number of births at the start of the life table against which the survivors at each age are compared. More generally it is the initial size of any cohort subjected to a particular chance of experiencing an event (Pressat, 1985).

In human demography a typical radix is assigned a number such as 100,000. Thus any number remaining at successive ages can be conveniently expressed as the number of survivors out of 100,000. In population biology the life table radix is usually assigned the value of unity, so that subsequent survivors are expressed as a fraction of the original number. The radix is associated with the survival column of the life table, denoted l_x. This gives the number of individuals surviving to age x. Historically the radix index is zero (i.e., newborns are aged zero days at the beginning of the interval).

Life Table Functions

The proportion of a cohort surviving from birth to exact age x is designated l_x. The difference in number of survivors for successive ages x and x + 1 is designated d_x:

$$d_x = l_x - l_{x+1} \qquad (2\text{-}1a)$$

and the difference in survivorship for ages n days apart is

$$_nd_x = l_x - l_{x+n} \qquad (2\text{-}1b)$$

Thus the d_x and $_nd_x$ schedules give the frequency distribution of deaths in the cohort.

The probability of surviving from age x to x + 1 is designated p_x, where

$$p_x = l_{x+1}/l_x \qquad (2\text{-}2a)$$

and more generally the probability of surviving from age x the age x + n is designated $_np_x$, where

$$_np_x = l_{x+n}/l_x \qquad (2\text{-}2b)$$

Both p_x and $_np_x$ are termed *period survivorship*.

The complements of these survival probabilities, designated $q_x(=1-p_x)$ and $_nq_x(=1-{_np_x})$ are termed *period mortality* and represent the probability of dying over these respective periods. Note that

$$q_x = \frac{d_x}{l_x} \qquad _nq_x = \frac{_nd_x}{l_x}$$

$$= \frac{l_x - l_{x+1}}{l_x} \qquad = \frac{l_x - l_{x+n}}{l_x}$$

$$= 1 - \frac{l_{x+1}}{l_x} \qquad = 1 - \frac{l_{x+n}}{l_x}$$

$$= 1 - p_x \qquad = 1 - {_np_x}$$

A concept fundamental to life table analysis is the number of days lived

LIFE TABLES

in an age interval. If we assume that individuals that die in an interval do so at the midpoint of the interval, then the number of days lived by the average individual in the cohort from x to x + 1, denoted L_x, is given by the formula

$$L_x = (l_x - d_x) + \tfrac{1}{2}d_x$$
$$= (l_x - (d_x/2))$$
$$= \frac{(l_x + l_{x+1})}{2} \quad (2\text{-}3a)$$

and for the age interval n-days apart is given as

$$_nL_x = n(l_x - \tfrac{1}{2}\,_nd_x)$$
$$= .5n(l_x + l_{x+n}) \quad (2\text{-}3b)$$

If L_x gives the number of days lived by the average individual within a cohort in the interval x to x + 1, then the total number of days to be lived by the average individual within a cohort from age x to the last day of possible life is

$$T_x = \sum_{y=x}^{\omega} L_y \quad (2\text{-}4)$$

where T_x denotes this total. Since there are l_x individuals that survive to age x in the cohort and a total of T_x insect-days remaining to these l_x individuals, the number of per capita days of life remaining to the average individual living at age x is

$$e_x = T_x/l_x \quad (2\text{-}5)$$

The term e_x denotes the expectation of life at age x. The average age of death of an individual age x is simply its current age plus the expectation of life at that age ($= x + e_x$).

Complete Cohort Life Table

A complete cohort life table is constructed by recording the number of deaths in an initial cohort of identically aged individuals at each point in time until all have died. The information is placed in columns according to the number alive at the beginning of each age interval, denoted K_x, and the number of deaths in the interval, denoted D_x. This information is then arranged by age class in eight columns representing the life table functions. An example of the data needed for life table construction and the complete cohort life table are given in Tables 2-1 and 2-2 for the human louse, *Pediculus humanus* (data from Evans and Smith, 1952).

The complete life table (Table 2-2) consists of eight columns:

Column 1

Age class. This is the age index and designates the exact age at which the interval begins *relative to the initial cohort*. Age class x includes the interval

Table 2-1. Life Table Construction for the Human Louse, *Pediculus humanus*, Using Data from Evans and Smith (1952) and a Hypothetical Cohort of 1000 Newborns

Stage	Age Class (Days), x	Age Interval (Days), x to x + 1	Number Alive at Beginning of Interval, K_x	Number of Deaths in Age Interval, D_x
Egg	0	0–1	1000	14
Egg	1	1–2	986	13
Egg	2	2–3	973	14
Egg	3	3–4	959	14
Egg	4	4–5	945	13
Egg	5	5–6	932	14
Egg	6	6–7	918	14
Egg	7	7–8	904	13
Egg	8	8–9	891	14
1st instar	9	9–10	877	7
1st instar	10	10–11	870	6
1st instar	11	11–12	864	7
1st instar	12	12–13	857	6
1st instar	13	13–14	851	7
2nd instar	14	14–15	844	6
2nd instar	15	15–16	838	6
2nd instar	16	16–17	832	5
2nd instar	17	17–18	827	4
3rd instar	18	18–19	823	8
3rd instar	19	19–20	815	8
3rd instar	20	20–21	807	7
3rd instar	21	21–22	800	7
Adult	22	22–23	793	8
Adult	23	23–24	785	2
Adult	24	24–25	783	20
Adult	25	25–26	763	12
Adult	26	26–27	751	17
Adult	27	27–28	734	24
Adult	28	28–29	710	14
Adult	29	29–30	696	26
Adult	30	30–31	670	32
Adult	31	31–32	638	31
Adult	32	32–33	607	28
Adult	33	33–34	579	44
Adult	34	34–35	535	43
Adult	35	35–36	492	32
Adult	36	36–37	460	44
Adult	37	37–38	416	43
Adult	38	38–39	373	30
Adult	39	39–40	343	30
Adult	40	40–41	313	33
Adult	41	41–42	280	6
Adult	42	42–43	274	28
Adult	43	43–44	246	20
Adult	44	44–45	226	28
Adult	45	45–46	198	28
Adult	46	46–47	170	23
Adult	47	47–48	147	18

LIFE TABLES

Table 2-1. (Contd.)

Stage	Age Class (Days), x	Age Interval (Days), x to x + 1	Number Alive at Beginning of Interval, K_x	Number of Deaths in Age Interval, D_x
Adult	48	48–49	129	20
Adult	49	49–50	109	26
Adult	50	50–51	83	18
Adult	51	51–52	65	21
Adult	52	52–53	44	8
Adult	53	53–54	36	14
Adult	54	54–55	22	12
Adult	55	55–56	10	1
Adult	56	56–57	9	1
Adult	57	57–58	8	1
Adult	58	58–59	7	1
Adult	59	59–60	6	1
Adult	60	60–61	5	1
Adult	61	61–62	4	1
Adult	62	62–63	3	1
Adult	63	63–64	2	1
Adult	64	64–65	1	1
Adult	65	65–66	0	—

Table 2-2. Complete Life Table for *Pediculus humanus*

Age Class, x (1)	Fraction Living at Age x, l_x (2)	Fraction Surviving from x to x + 1, p_x (3)	Fraction Dying from x to x + 1, q_x (4)	Fraction Dying in Interval x to x + 1, d_x (5)	Days Lived in Interval, L_x (6)	Number Days Lived Beyond Age x, T_x (7)	Expectation of Life, e_x (8)
0	1.000	.986	.014	.014	.993	32.438	32.438
1	.986	.987	.013	.013	.980	31.445	31.891
2	.973	.986	.014	.014	.966	30.465	31.311
3	.959	.985	.015	.014	.952	29.499	30.761
4	.945	.986	.014	.013	.939	28.548	30.209
5	.932	.985	.015	.014	.925	27.609	29.623
6	.918	.985	.015	.014	.911	26.684	29.068
7	.904	.986	.014	.013	.898	25.773	28.510
8	.891	.984	.016	.014	.884	24.876	27.919
9	.877	.992	.008	.007	.874	23.992	27.356
10	.870	.993	.007	.006	.867	23.118	26.572
11	.864	.992	.008	.007	.861	22.251	25.753
12	.857	.993	.007	.006	.854	21.391	24.960
13	.851	.992	.008	.007	.847	20.537	24.132
14	.844	.993	.007	.006	.841	19.689	24.328
15	.838	.993	.007	.006	.835	18.848	22.492
16	.832	.994	.006	.005	.830	18.013	21.650
17	.827	.995	.005	.004	.825	17.184	20.778
18	.823	.990	.010	.008	.819	16.359	19.877
19	.815	.990	.010	.008	.811	15.540	19.067
20	.807	.991	.009	.007	.803	14.729	18.251

Table 2-2. (Contd.)

Age Class, x (1)	Fraction Living at Age x, l_x (2)	Fraction Surviving from x to x + 1, p_x (3)	Fraction Dying from x to x + 1, q_x (4)	Fraction Dying in Interval x to x + 1, d_x (5)	Days Lived in Interval, L_x (6)	Number Days Lived Beyond Age x, T_x (7)	Expectation of Life, e_x (8)
21	.800	.991	.009	.007	.796	13.925	17.406
22	.793	.990	.010	.008	.789	13.128	16.555
23	.785	.997	.003	.002	.784	12.340	15.719
24	.783	.974	.026	.020	.773	11.555	14.758
25	.763	.984	.016	.012	.757	10.783	14.132
26	.751	.977	.023	.017	.742	10.026	13.350
27	.734	.967	.033	.024	.722	9.283	12.647
28	.710	.980	.020	.014	.703	8.561	12.058
29	.696	.963	.037	.026	.683	7.858	11.290
30	.670	.952	.048	.032	.654	7.175	10.709
31	.638	.951	.049	.031	.623	6.521	10.221
32	.607	.954	.046	.028	.593	5.899	9.717
33	.579	.924	.076	.044	.557	5.306	9.163
34	.535	.920	.080	.043	.514	4.749	8.876
35	.492	.935	.065	.032	.476	4.235	8.608
36	.460	.904	.096	.044	.438	3.759	8.172
37	.416	.897	.103	.043	.395	3.321	7.983
38	.373	.920	.080	.030	.358	2.927	7.846
39	.343	.913	.087	.030	.328	2.568	7.488
40	.313	.895	.105	.033	.297	2.241	7.158
41	.280	.979	.021	.006	.277	1.944	6.943
42	.274	.898	.102	.028	.260	1.667	6.084
43	.246	.919	.081	.020	.236	1.407	5.720
44	.226	.876	.124	.028	.212	1.171	5.181
45	.198	.859	.141	.028	.184	0.959	4.843
46	.170	.865	.135	.023	.159	0.775	4.559
47	.147	.878	.122	.018	.138	0.616	4.194
48	.129	.845	.155	.020	.119	0.479	3.709
49	.109	.761	.239	.026	.096	0.360	3.298
50	.083	.783	.217	.018	.074	0.264	3.175
51	.065	.677	.323	.021	.055	0.189	2.915
52	.044	.818	.182	.008	.040	0.135	3.068
53	.036	.611	.389	.014	.029	0.095	2.639
54	.022	.455	.545	.012	.016	0.066	3.000
55	.010	.900	.100	.001	.010	0.050	5.000
56	.009	.889	.111	.001	.008	0.041	4.500
57	.008	.875	.125	.001	.008	0.032	4.000
58	.007	.857	.143	.001	.007	0.024	3.500
59	.006	.833	.167	.001	.006	0.018	3.000
60	.005	.800	.200	.001	.005	0.012	2.500
61	.004	.750	.250	.001	.004	0.008	2.000
62	.003	.667	.333	.001	.003	0.005	1.500
63	.002	.500	.500	.001	.002	0.002	1.000
64	.001	.000	1.000	.001	.001	0.001	0.500
65	.000	—	—	.000			
				1.000			

LIFE TABLES

from exact age x to exact age x + 1. For example, age class 0 specifies the interval from age 0 to age 1.

Column 2
Fraction of the original cohort alive at age x, l_x. The first fraction in this column, l_0, is the radix, and each successive number represents the fraction of survivors at the exact age x from the cohort of size l_0 (normalized to 1.0). For example, 807 individuals in the cohort survived to age 20. Thus .807 of the original cohort survived to this age since the cohort stated with 1,000 newborn.

Column 3
Proportion of those alive at age x that survive through the interval x to x + 1, p_x. For example,

$$p_0 = l_1/l_0 \qquad p_1 = l_2/l_1$$
$$= .986/1.000 \qquad = .973/.986$$
$$= .986 \qquad = .987$$
$$p_{50} = l_{51}/l_{50} \qquad p_{51} = l_{52}/l_{51}$$
$$= .065/.083 \qquad = .044/.065$$
$$= .783 \qquad = .677$$

Column 4
Proportion of those alive at age x that die in the interval x to x + 1, q_x, For example,

$$q_0 = 1.000 - p_1 \qquad q_1 = 1.000 - p_2$$
$$= 1.000 - .986 \qquad = 1.000 - .987$$
$$= .014 \qquad = .013$$
$$q_{50} = 1.000 - p_{51} \qquad q_{51} = 1.000 - p_{52}$$
$$= 1.000 - .783 \qquad = 1.000 - .677$$
$$= .217 \qquad = .323$$

Column 5
Fraction of the original cohort, l_0, that die in the age interval x to x + 1, d_x. Therefore, the d_x column represents the frequency distribution of deaths in the cohort and its sum is unity.

$$d_0 = l_0 - l_1 \qquad d_1 = l_1 - l_2$$
$$= 1.000 - .986 \qquad = .986 - .973$$
$$= .014 \qquad = .013$$
$$d_{50} = l_{50} - l_{51} \qquad d_{51} = l_{51} - l_{52}$$
$$= .083 - .065 \qquad = .065 - .044$$
$$= .018 \qquad = .021$$

Column 6
Per capita fraction of interval lived in the age interval x to x+1, L_x. For example,

$$L_0 = l_0 - (1/2)d_0 \qquad L_1 = l_1 - (1/2)d_1$$
$$= 1.000 - (1/2)(.014) \qquad = .986 - (1/2)(.013)$$
$$= .993 \qquad = .980$$
$$L_{50} = l_{50} - (1/2)d_{50} \qquad L_{51} = l_{51} - (1/2)d_{51}$$
$$= .083 - (1/2)(.018) \qquad = .065 - (1/2)(.021)$$
$$= .074 \qquad = .055$$

Column 7
Total number of days lived beyond age x, T_x. This total is essential to the computation of life expectancy since it gives the number of insect-days lived by the cohort after age x uncorrected for the total beginning at age x. For example,

$$T_0 = \sum_{x=0}^{\omega} L_x \qquad T_1 = \sum_{x=1}^{\omega} L_x$$
$$= L_0 + L_1 + \cdots + L_{63} + L_{64} \qquad = L_1 + L_2 + \cdots + L_{63} + L_{64}$$
$$= .993 + .980 + \cdots + .002 + .001 \qquad = .980 + .966 + \cdots + .022 + .001$$
$$= 32.438 \qquad = 31.445$$
$$T_{50} = \sum_{x=50}^{\omega} L_x \qquad T_{51} = \sum_{x=51}^{\omega} L_x$$
$$= L_{50} + L_{51} + \cdots + L_{63} + L_{64} \qquad = L_{51} + L_{52} + \cdots + L_{63} + L_{64}$$
$$= .074 + .055 + \cdots + .002 + .001 \qquad = .055 + .040 + \cdots + .022 + .001$$
$$= .264 \qquad = .189$$

Column 8
Expectation of life at age x, e_x. This gives the average remaining lifetime for an individual who survives to the beginning of the indicated age interval. For example,

$$e_0 = T_0/l_0 \qquad e_1 = T_1/l_1$$
$$= 32.438/1.000 \qquad = 31.445/.986$$
$$= 32.438 \qquad = 31.891$$
$$e_{50} = T_{50}/l_{50} \qquad e_{51} = T_{51}/l_{51}$$
$$= .264/.083 \qquad = .189/.065$$
$$= 3.175 \qquad = 2.915$$

A number of relationships emerge from this life table analysis that merit comment. *First*, the expectation of life for a newborn louse is over one month. A louse that survives to one month is expected to live an average of 10 days more. A two-month-old (60 days) louse lives an average of only 2 more days.

LIFE TABLES

Second, around a third of all deaths occur in the first 30 days, but another third of all deaths occur in the following 10 days (i.e., 30 to 40 days). The last third of all deaths occur over the last 3 weeks of possible life. *Third*, the probability of surviving from age 0 to 35 days is around .50. However, of all those alive at age 36 the probability of surviving for the next 8 days is also .50. *Fourth*, the probability of dying from age x to x + 1 when lice are under 10 days old is up to 50-fold less than when they are over 50 days old. For example, the probability of an individual's dying from age 54 to 55 days is 54-fold greater than the probability of the same individual's dying from age 19 to 20 days.

THE ABRIDGED LIFE TABLE

The complete life table has two disadvantages that can be removed by constructing an abridged table. *First*, it is sometimes not possible to monitor daily mortality of individuals in a cohort over their entire life course. For example, it is extremely difficult to determine the precise time of death for eggs and pupae for most insects. Thus the practical solution for determining cohort mortality for these stages is to note the number entering and the number surviving through the stage, which will yield period (stage) survival. Because the duration of each stage is typically greater than one day and two stages seldom have the same duration, the methodology for constructing the complete life table is not appropriate. *Second*, a table of 50 to 100 age groups with 5 to 7 life table functions (columns) is difficult to fully comprehend and contains details that are often not of concern. By grouping deaths into larger intervals it is possible to summarize the information concisely while still retaining the basic life table format and concepts.

The abridged life table generally contains the same functions as the complete table. In addition, the duration of each stage is given as n to specify the age interval over which mortality is assessed. An example of an abridged life table is given in Table 2-3 for worker honey bees, *Apis mellifera*.

Column 1
Stage and duration (n). This gives the stage over which mortality is measured and the duration of the stage. Preadult stages can be further subdivided into instars, and adults can be divided into various physiological divisions such as preovipositional and ovipositional or into arbitrary age groupings as given for the honey bee.

Column 2
Age index, x. This column gives the age associated with each stage. For example, the egg stage lasts 3 days, beginning at age $x = 0$ and extending to age $x + n = 0 + 3 = 3$. This is the starting age index for the unsealed brood.

Table 2-3. Abridged Life Table for Worker Honey Bees (Data from Sakagami and Fukuda, 1986)

STAGE (n = duration in days) (1)	Age Interval, x (2)	Parameter						
		l_x (3)	$_np_x$ (4)	$_nq_x$ (5)	$_nd_x$ (6)	$_nL_x$ (7)	T_x (8)	e_x (9)
Egg (n = 3)	0–3	1.000	.958	.042	.042	2.937	40.482	40.482
Unsealed brood (n = 5)	3–8	.958	.857	.143	.137	4.448	37.545	39.191
Sealed brood (n = 12)	8–20	.821	.988	.012	.010	9.792	33.097	40.313
Adult (n = 10)	20–30	.811	.945	.055	.031	7.955	23.305	28.736
Adult (n = 10)	30–40	.780	.947	.053	.041	7.595	15.350	19.679
Adult (n = 10)	40–50	.739	.499	.501	.370	5.540	7.755	10.494
Adult (n = 10)	50–60	.369	.100	.900	.332	2.030	2.215	6.003
Adult (n = 10)	>60	.037	.000	1.00	.037	.185	.185	5.000
					1.000	40.482		

The unsealed brood lasts 5 days (n = 5) beginning at age x = 3 and extending to age x + n = 3 + 5 = 8. The adult stage is divided into 10-day increments beginning at age 20 days when the average preadult matures.

Column 3
Fraction of the original cohort alive at the beginning of the designated age interval, x to x + n, l_x. This measure corresponds exactly to the l_x column in the complete life table. For example, the fraction that survives to the unsealed brood is .958, and to the adult stage it is .811.

Column 4
Proportion of those alive at age x that survive through the interval x to x + n, $_np_x$. For example,

$$_3p_0 = l_3/l_0 (= \text{egg survival})$$
$$= .958/1.000$$
$$= .958$$
$$_{12}p_8 = l_{20}/l_8 (= \text{sealed brood survival})$$
$$= .811/.821$$
$$= .988$$

Column 5
Proportion of those alive at age x that die in the interval x to x + n, $_nq_x$. For example,

$$_3q_0 = 1 - {_3p_0} (= \text{egg mortality})$$
$$= 1 - .958$$
$$= .042$$

LIFE TABLES

$$_{12}q_8 = 1 - {_{12}p_8}(= \text{sealed brood mortality})$$
$$= 1 - .988$$
$$= .012$$

Column 6
Fraction of the original cohort, l_0, that die in the age interval x to x + n, $_nd_x$. For example,

$$_3d_0 = l_0 - l_3 \text{ (fraction of all deaths in egg)}$$
$$= 1.000 - .958$$
$$= .042$$
$$_{12}d_8 = l_8 - l_{20}(\text{fraction of all deaths in sealed brood})$$
$$= .821 - .811$$
$$= .010$$

Column 7
Per capita fraction of interval lived in the age interval x to x + n, $_nL_x$. For example,

$$_3L_0 = 3[l_0 - (1/2)_3d_0]$$
$$= 3[1.000 - (1/2).042]$$
$$= (3)(.979)$$
$$= 2.937$$
$$_{12}L_8 = 12[l_8 - (1/2)_{12}d_8]$$
$$= 12[.821 - (1/2).010]$$
$$= (12)(.816)$$
$$= 9.792$$

Column 8
Total number of days lived beyond age x, T_x. For example,

$$T_0 = {_3L_0} + {_5L_3} + {_{12}L_8} + {_{10}L_{20}} + \cdots + {_{10}L_{60}}$$
$$= 2.937 + 4.448 + 9.792 + 7.955 + \cdots + .185$$
$$= 40.482$$
$$T_8 = {_{12}L_8} + {_{10}L_{20}} + \cdots + {_{10}L_{60}}$$
$$= 9.792 + 7.955 + \cdots + .185$$
$$= 33.097$$

Column 9
Expected number of additional days the average individual age x will live, e_x. For example,

$$e_0 = T_0/l_0$$

$$= 40.482/1.000$$
$$= 40.482$$
$$e_8 = T_8/l_8$$
$$= 33.097/.821$$
$$= 40.313$$

Several aspects of the mortality and survival of worker bees are noteworthy (i) over 80% of all newborn survive to adulthood and over one-third survive to age 50 days (from l_x schedule); (ii) the probability of dying from age 20 to 30 (first 10 days of adulthood) is around 10-fold less than in the interval from 40 to 50 days old (i.e., when adults are 20 to 30 days old); and (iii) the expectation of life of a newly enclosed adult worker beee is nearly one month.

THE MULTIPLE DECREMENT LIFE TABLE

The multiple decrement life table is used widely in human actuarial studies to address questions concerning the frequency of occurrence for causes of death and how life expectancy might change if certain causes were eliminated. The conventional single decrement life table shows the probability of survivorship of an individual subject to the one undifferentiated hazard of death. In multiple decrement tables the individual is subject to a number of mutually exclusive hazards, such as disease, predators, or parasites, and is followed in the table only to its exit, as in the ordinary life table. But in the multiple decrement table there is now more than one way of exiting (Anon., 1962; Preston et al., 1972; Carey, 1989).

Two probabilities and hence two kinds of tables are commonly recognized in the study of cause of death. One is the probability of dying of a certain cause in the presence of other causes; the other is the probability of dying of a certain cause in the absence of other causes (Preston et al., 1972). The first gives rise to the multiple decrement table proper. The second gives rise to an associated single decrement table and is applied to find the probability of dying if one or more factors were to disappear as a cause of death.

The assumption of the multiple decrement life table is that multiple causes of death act independently. It is concerned with the probability that an individual will die of a certain cause in the presence of other causes. The concept itself stems from reliability theory in operations research. Keyfitz (1982, 1985) uses an example of a watch that can operate only as long as all its parts are functioning: each part has its own life table. The probability that an individual (i.e., the watch) will survive to a given age is the product of the independent probabilities that each of its components will "survive" to that age. The same notion of probabilities applied to internal components causing the death of a system can also be applied to external components such as disease and accidents in humans or predation and parasitism in insects. The concept here is that the probability of an insect's surviving to a certain age (or stage) is the product of all independent risk probabilities.

LIFE TABLES

In general, multiple decrement theory is basically concerned with three questions (Elandt-Johnson and Johnson, 1980): i) What is the age (stage) distribution of deaths from different causes acting simultaneously in a given population? ii) What is the probability that a newborn individual will die after a given age or stage from a specified cause? iii) How might the mortality pattern or expectation of life change if certain causes were eliminated? The first two questions are concerned with evaluating patterns and rates of mortality, while the last question is concerned with what is termed "competing risk analysis." In both cases the analyses are based on three assumptions: i) each death is due to a single cause; ii) each individual in a population has exactly the same probability of dying from any of the causes operating in the population (see Moriyama, 1956; Vaupel and Yashin, 1984); and iii) the probability of dying from any given cause is independent of the probability of dying from any other source.

Data and Data Organization

A hypothetical data set for analyzing mortality in a synthetic cohort was derived using average stage-by-cause mortality from 25 life tables given in Cameron and Morrison (1977) for the apple maggot, *Rhagoletis pomonella*. The original data were divided into death due to 11 factors—one for egg, four for larval, and six for pupal and adult emergence. These sources of mortality are here lumped into four categories: predation, parasitism, disease, and other causes. The group within which a cause of death was placed was arbitrary in several cases.

The hypothetical mortality data for the four categories (causes) of death in preadult *R. pomonella* are given in Table 2-4, where K_x is the number in the cohort aged x, D_x is the total number of deaths in stage x, and D_{ix} is the number of deaths due to cause i in stage x. Note that $D_x = D_{1x} + D_{2x} + D_{3x} + D_{4x}$ and also that the K_x column does not represent the non-normalized survival column. That is, the K_x column gives the number of insects at the

Table 2-4. Deaths from Four Causes in *Rhagoletis pomonella* Populations Using a Hypothetical Data Set[1].

Stage (index), x	Number Beginning Stage, K_x	Total Deaths, D_x	Number Deaths from—			
			Predators, D_{1x}	Parasites, D_{2x}	Disease, D_{3x}	Other causes, D_{4x}
Egg (1)	977	14	0	0	0	14
Early larvae (2)	963	810	0	224	0	586
Late larvae (3)	153	112	100	12	0	0
Early pupae (4)	435	98	88	0	10	0
Late pupae (5)	351	206	133	0	19	54
Adult (6)	—	—	—	—	—	—

[1] The data presented in Cameron and Morrison (1977) were used as guidelines for the relative numbers of deaths by cause.

beginning of the stage that were exposed to risk through the stage. This is not the same as the number that would be exposed to risk in a true cohort where the numbers would decrease from stage to stage.

General Framework and Notation

The notation for all functions in the multiple decrement table corresponds to the single decrement cases except i) the prefix a is added to denote "in presence of all causes"; and ii) the symbol x is used to denote the stage index rather than the age interval. Therefore, let

al_{ix} = fraction of original cohort living at age x that ultimately die from cause i

al_x = fraction of survivors at age x out of original cohort of al_1 (start index at x = 1)

ad_{ix} = fraction of deaths in stage x from cause i among al_x living at stage x

ad_x = fraction of deaths in stage x from all causes ($= ad_{1x} + ad_{2x} + \cdots + ad_{kx}$)

aq_{ix} = fraction of deaths from cause i in stage x in the presence of all other causes, given that the individual is alive at beginning of stage x

aq_x = fraction of deaths from all causes in stage x, given that individual is alive at stage x ($= aq_{1x} + aq_{2x} + \cdots + aq_{kx}$)

The fraction dying in the interval designated aq_x is

$$aq_x = D_x/K_x \tag{2-6}$$

The fraction of the cohort age x dying in stage x due to cause i is given by

$$aq_{ix} = D_{ix}/K_x \tag{2-7}$$

For example, no deaths occurred in the egg stage due to predators, parasites, or disease. Thus

$$aq_{1,1} = aq_{2,1} = aq_{3,1} = 0.0$$

However, 14 of 977 eggs died of "other causes", therefore

$$aq_{4,1} = D_{4,1}/K_1$$
$$= 14/977$$
$$= .01433$$

and

$$aq_1 = aq_{4,1} = .01433$$

since "other causes" was the only source of death. The complete table of death probabilities based on the mortality data of Table 2-4 is given in Table 2-5. The computation of these rates is necessary for completing the full

LIFE TABLES

Table 2-5. Cause-Specific Probability of Death from Specified Causes in the Presence of All Causes for *Rhagoletis pomonella* using Hypothetical Data Presented in Table 2-4

Stage (Index), x	Total, aq_x	Cause of Death			
		Predators, aq_{1x}	Parasites, aq_{2x}	Disease, aq_{3x}	Other, aq_{ux}
Egg (1)	.01433	.00000	.00000	.00000	.01433
Early larvae (2)	.84112	.00000	.23261	.000000	.60851
Late larvae (3)	.73203	.65359	.07842	.00000	.00000
Early pupae (4)	.22529	.20230	.00000	.02299	.00000
Late pupae (5)	.58690	.37892	.00000	.05413	.15385
Adult (6)	1.00000	—	—	—	—

multiple decrement analysis. Note in Table 2-5 that the stage- and cause-specific mortality rates derived from the data are now expressed as per capita probabilities. Two aspects of this table may be noted: i) the highest death rate is due to late larval predation; and ii) the highest stage-specific mortality occurs in early larvae and is due to both predation and other causes.

Table Construction

The main multiple decrement table uses the aq_{ix} value in Table 2-5 to determine schedules for the fraction of the starting cohort dying in stage x from cause i ($ad_{i,x}$), the total fraction dying in stage x from all causes (ad_x), and the fraction of newborn surviving to stage x (al_x). These are computed as follows:

Step 1. Compute survival to stage x subject to all causes. We use an index of x = 1 for the first stage (i.e., egg), set $al_1 = 1.0$, and compute progressively

$$al_{x+1} = al_x(1.0 - aq_x)$$

For example,

$$al_2 = al_1(1.0 - aq_1)$$
$$= 1.0(1.0 - .0143)$$
$$= .9857$$
$$al_3 = .9857(1.0 - .8411)$$
$$= .1566$$

Step 2. Compute the fraction of newborns dying in stage x from all causes. This is computed as

$$ad_x = al_x - al_{x+1}$$

For example,

$$ad_1 = al_1 - al_2$$
$$= 1.0 - .9857$$
$$= .0143$$

Step 3. Compute the fraction of newborns dying in stage x from cause i. We use the formula

$$ad_{i,x} = al_x(aq_{i,x})$$

For example,

$$ad_{4,1} = al_1(aq_{4,1})$$
$$= 1.0(.0143)$$
$$= .0143$$
$$ad_{1,3} = al_3(aq_{1,3})$$
$$= .1566(.65359)$$
$$= .1024$$

Values for the various relationships are given in Table 2-6. This table reveals relations that were not evident from Table 2-4. For example, nearly 83% of all deaths occurred in the early larval stage, only about 4% of all newborn survived to the pupal stage, nearly 62% of all deaths were a result of other causes, and disease accounted for less than 1% of all deaths.

Elimination of Cause—Concept

Farr (1875) apparently was the first to ask the question, what would be the effect on life expectancy if a certain disease were eliminated as a cause of death? This question is particularly germane to management questions since if it is possible to gain an understanding of the effect on life expectancy of eliminating a particular source of death, it follows that the same methods could be used to determine the impact of adding a source of death.

The only definitive method for determining the effect on expectation of life in arthropod populations of eliminating a certain cause of death is through experiment (DeBach and Huffacker, 1971; Royama, 1984; Luck et al., 1988). However, experimentation is sometimes either not possible or the only data available is natural history. Therefore it is necessary mathematically to approximate the effect of eliminating a certain cause. One such approach is described as follows. Suppose the probability of surviving factor A alone is p_A and the probability of surviving factor B alone is p_B. Then the probability of surviving both independent causes together, denoted p_{AB}, is given as

$$p_{AB} = p_A p_B \qquad (2\text{-}8a)$$

or

$$p_{AB} = (1 - q_A)(1 - q_B) \qquad (2\text{-}8b)$$

where q_A and q_B are complements of p_A and p_B, respectively. If D_A and D_B denote the fraction of all individuals observed that died of cause A and B, respectively, then

$$p_{AB} = 1 - (D_A + D_B) \qquad (2\text{-}9a)$$

and

$$1 - (D_A + D_B) = (1 - q_A)(1 - q_B) \qquad (2\text{-}9b)$$

Table 2-6. The Multiple Decrement Life Table for Life Table Deaths from Specified Causes at Given Stage of *R. pomonella*

Stage (Index), x	Probability of Death, aq_x	Fraction Living at Beginning of Interval, al_x	Fraction of All Deaths, ad_x	Fraction Deaths from —			
				Predator, ad_{1x}	Parasites, ad_{2x}	Disease, ad_{3x}	Other Causes, ad_{4x}
Egg (1)	.0143	1.0000	.0143	.0000	.0000	.0000	.0143
Early larvae (2)	.8411	.9857	.8291	.0000	.2293	.0000	.5998
Late larvae (3)	.7320	.1566	.1146	.1024	.0122	.0000	.0000
Early pupae (4)	.2253	.0420	.0095	.0085	.0000	.0010	.0000
Late pupae (5)	.5869	.0325	.0191	.0123	.0000	.0018	.0050
Adult (6)	—	.0134	—	—	—	—	—
Total[1]	—		.9866	.1232	.2415	.0028	.6191

[1] Apply only to the total mortality to the adult stage (i.e., ad_x) and the totals for each of the four mortality causes (i.e., ad_{ix}'s)

The objective is to obtain values for q_A and q_B since we would like to determine mortality in the absence of one or the other factor. It is necessary to specify a second equation since Equation 2-9b has two unknowns. By assuming that the ratio of numbers dying from factor A to the numbers dying from factor B equals the ratio of the probability of dying from factor A to the probability of dying from factor B, we can obtain the second equation:

$$\frac{q_A}{q_B} = \frac{D_A}{D_B} \qquad (2\text{-}10)$$

Therefore Equations 2-9b and 2-10 represent two simultaneous equations in two unknowns (q_A and q_B). Expressing q_A in Equation 2-10 in terms of q_B, D_A, and D_B and then substituting this expression in Equation 2-9b yields the quadratic equation

$$aq_B^2 + bq_B + c = 0 \qquad (2\text{-}11)$$

where $a = D_A$; $b = -(D_A + D_B)$; $c = D_B(D_A + D_B)$.

The value of q_B is found by substituting a, b, and c into the quadratic formula

$$q_B = \frac{-b - \sqrt{b^2 - 4ac}}{2a} \qquad (2\text{-}12)$$

Elandt-Johnson and Johnson (1980), Namboordiri and Suchindran (1987), and Preston et al. (1972) present alternative approaches for finding the solutions to the independent risk probabilities.

As an example, suppose that of 1000 individuals observed over their preadult lifetime, 20 remained alive (i.e., 2%), 370 died as a result of natural enemies and 610 died of other causes. Therefore set $D_A = .37$ and $D_B = .61$. Substituting these values into Equation 2-12 yields $a = .37$, $b = -.98$, and $c = .598$. Therefore,

$$q_B = \frac{[.98 - \sqrt{(.98)^2 - 4(.37)(.598)}]}{2(.37)}$$

$$= .953$$

and

$$q_A = q_B D_A / D_B$$
$$= (.953)(.37)/.61$$
$$= .578$$

A graphical interpretation of the analysis is presented in Figure 2-1. The results state that if factor A were completely eliminated as a source of mortality, factor B alone would still kill 95.3% of the original cohort. This is a substantial increase from the 61% mortality owing to this factor in the presence of factor A. On the other hand, 57.8% would die if factor A alone accounted for deaths. In short, adding factor A as a cause of mortality when B is already present would increase mortality from 95.3% to 98%, or less than 3%. However, adding factor B as a cause of mortality when factor A

LIFE TABLES

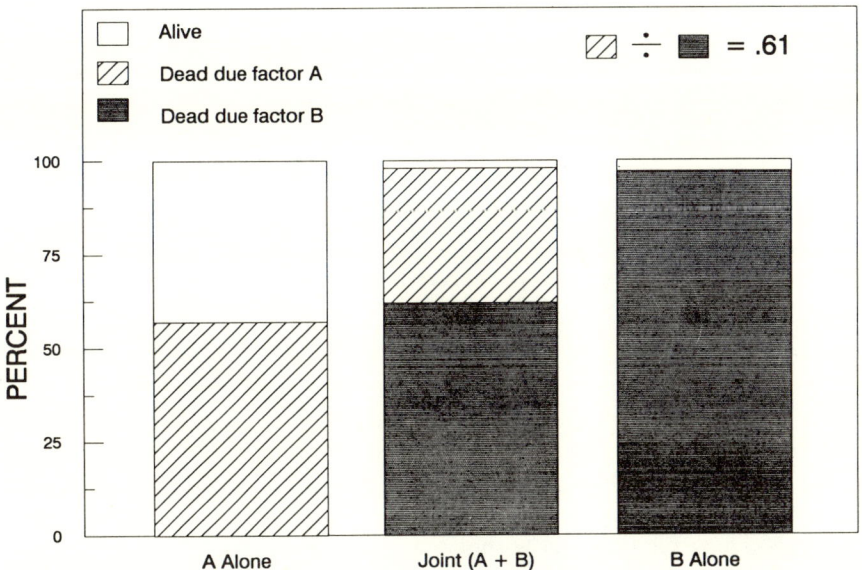

Figure 2-1. Illustration of the concept of competing risk in hypothetical cohorts subject to mortality factors A alone, B alone, and A and B jointly. Note that the ratio of those dead owing to A alone (q_A) and B alone (q_B) equals the ratio of the cause-specific mortalities when the two factors act jointly (i.e., D_A & D_B). That is, $q_A/q_B = D_A/D_B = .61$ (redrawn from Carey, 1989).

is already present would increase mortality from 57.8% to 98% or by over 40%. These differences are referred to in the ecological literature as indispensable (or irreplaceable) mortality (Huffaker and Kennet, 1965; Southwood, 1971).

Note that computationally the concept of double decrement given here embraces all the issues of multiple decrement (Preston et al., 1972). That is, no matter how many causes are considered, the probability of dying from each can be computed by considering the one in question versus "all others".

Elimination of Cause—Application

The data presented in Table 2-4 are used to compute the independent stage-by-cause probabilities of dying, q_{ix}'s. Note from this table that the egg stage has a single risk ("Other Causes"); early larvae, late larvae, and early pupae have two competing risks each, and late pupae have three competing risks each. Thus for the egg stage we have

$$q_{11} = q_{21} = q_{31} = 0$$

and

$$q_{41} = 1 - (963/977)$$
$$= .0143$$

The independent probabilities for the two competing risks in each of the next three stages can be computed using the same relationships described in the earlier example with the quadratic equation. For example, 224 of 963 individuals were parasitized in stage 2 (early larvae) and 586 died of other causes. Therefore, let D_A denote the fraction of the total that died of parasites,

$$D_A = 224/963 = .2326$$

and let D_B denote the fraction that died of other causes,

$$D_B = 586/963 = .6085$$

Substituting these values in Equation 2-12 yields values for $q_{22} = .296$ and $q_{42} = .775$. Probabilities for the three causes in stage 5 (i.e., predators, disease, & other causes) are determined by applying the quadratic to three 2-cause cases: i) predators vs. (disease + other causes); ii) disease vs. (predators + other causes); and iii) other causes vs. (predators + disease).

The results of this analysis are given in Table 2-7, where the q_{ix}'s denote the probability of dying in stage x of cause i in the absence of all other causes of death. These relations show that if a single cause of death were retained in the population, predation alone would reduce the cohort by over 87%, while other causes would reduce it by around 76%. This ranking of importance differs from the mortality data in the presence of all causes given in Table 2-5. From this perspective, predators appear to be less important in the presence of all causes since they attack later stages. Thus earlier causes of death reduced the number at risk in the stages susceptable to predation.

The q_{xi} values in Table 2-7 were used to compute the effect of various combinations of factors on total mortality. For example, the effect of predators + parasites on total mortality was computed by i) determining the probability of surviving each source in the absence of other sources over all stages—i.e., $(1-q_{1x})$ and $(1-q_{2x})$; ii) obtaining the product of the two survival probabilities within each stage; and iii) computing the product of these products over all stages. Since this gives total survival, 1 minus this value yields total mortality. The computations for this two-cause case are $[(1-0)(1-0)] \times [(1-0)(1-.296)] \times [(1-.708)(1-.085)] \times$

Table 2-7. Stage- and Cause-Specific Probability of Death for R. pomonella in the Absence of All Other Causes (Totals computed prior to rounding)

| Stage (Index), x | Total, q_x | Cause of Death ||||
		Predators, q_{1x}	Parasites, q_{2x}	Disease, q_{3x}	Other, q_{4x}
Egg (1)	.014	.000	.000	.000	.014
Early larvae (2)	.842	.000	.296	.000	.775
Late larvae (3)	.733	.708	.085	.000	.000
Early pupae (4)	.225	.207	.000	.023	.000
Late pupae (5)	.587	.458	.000	.065	.186
Egg to adult	.987	.874	.356	.087	.762

LIFE TABLES

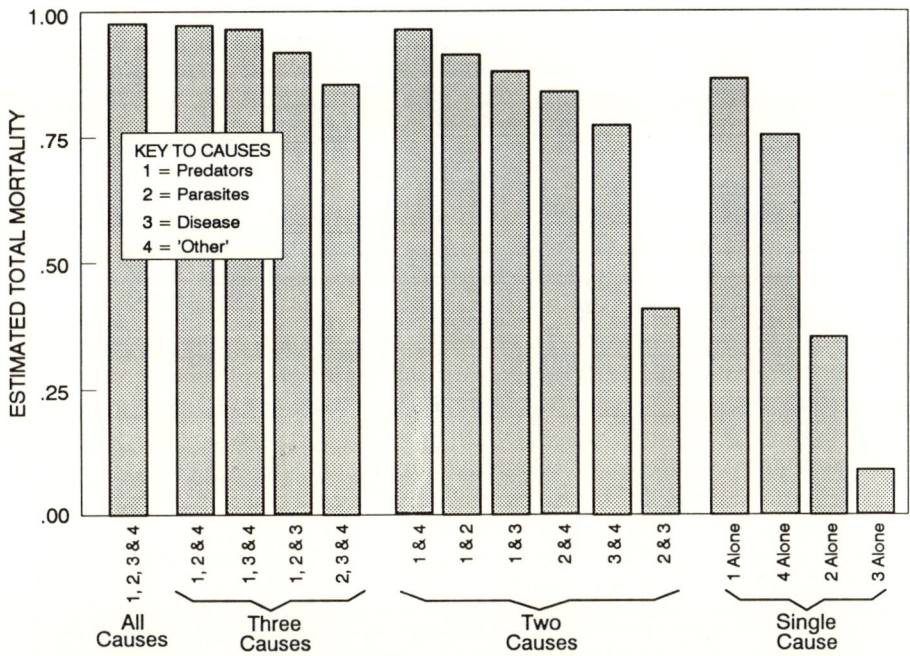

Figure 2-2. Estimated total mortality in *R. pomonella* cohorts for different cause-specific combinations.

$[(1 - .207)(1 - 0)] \times [(1 - .458)(1 - 0)] = .081$. This represents the fraction surviving to adulthood. Thus $(1 - .081) = .919$ gives the total preadult mortality.

The results of the complete analysis are given in Figure 2-2. Several aspects of these results merit comment. *First*, the effect of simultaneously eliminating multiple cause-of-death agents from the population cannot be inferred from observing the effect of eliminating each individually. The total contribution to mortality exceeds the sum of the individual components of total mortality (Preston et al., 1972). *Second*, while predators alone would kill 87% of an original cohort, predators plus other causes would reduce the population by nearly 98%. Thus the effect of adding parasites and disease to the system in the presence of the other two factors is negligible. *Third*, any pairwise combination of parasites, disease, and other causes, as well as all three causes combined, would reduce the population less than would predation alone. Conversely, adding or subtracting predation as a source of mortality affects total mortality much greater than adding or subtracting any other single source.

General Concepts

Multiple decrement theory embraces most of the major concepts and techniques currently used in insect mortality analysis, including the conventional life table (Deevey, 1947), Abbott's correction (Abbott, 1925), key factor analysis (Varley, 1947; Morris, 1965; Harcourt, 1969), tests for joint chemical toxicity (e.g., Hewlett and Plackett, 1959), and probit analysis (Finney, 1964). The interconnections of these tools and multiple decrement theory are described as follows:

1. The independent variable for each is time, age, stage, or dose. And for chemical tests such as probit analysis, dose and age are interchangeable in that age can be viewed as a dose of time. Thus probit analysis and life table analysis are conceptually and statistically identical (see Carey, 1986). Abbott's correction is simply a double decrement, single-time-step life table and tests of joint toxicity are essentially multiple decrement, one-time-step life tables. Key factor analysis orders events by stage as well as causes of death within a stage and is therefore a type of sequential risk life table.
2. All of the techniques either explicitly or implicitly rely on the assumption of competing risk. That is, the removal of one of several mortality factors within a stage or prior to the stage will change the number of individuals exposed to the risk of the cause in question. Thompson (1955) labeled these contemporaneous mortality factors, and Huffaker and Kennett (1966) referred to the non-additive changes in mortality on removal of one of several competing risks as compensatory mortality. In the conventional single decrement life table, death at early stages is "competing" against death at later stages. That is, a small fraction of the total deaths occur at older ages simply because most individuals die before attaining old age. This is a form of sequential competing risk.
3. Most of the techniques are based on the assumption of independence. In conventional life table analysis it is assumed that the probability of surviving from age x to age $x + 1$ is independent of the probability of surviving from age $x - 1$ to age x and also independent of density. This assumption is explicit in Abbott's correction or in the tests for joint toxicity where one chemical does not change the biological effect in the presence of another chemical. Likewise, in multiple decrement life tables it is assumed, for example, that if an insect is infected with a pathogen it is no more susceptible to predation than if it were not infected and that the associated probabilities of dying of either are density independent. Although it is commonly understood that few mortality factors are totally independent of the presence of other factors or of density, efforts at measuring and modeling these aspects have been less than satisfactory.

Despite the fact that demographers concerned with humans originated the life table that was introduced to the ecological literature by Deevey (1947) and is now viewed as convention, ecologists have subsequently resisted drawing from the human demographic and actuarial literature for analytical techniques concerned with survival. The reason most frequently given for this provincialism concerns differences in the quality of the data. While the data on causes of death and death rates in plant and animal populations may be less accurate than human vital statistics, concepts for data evaluation are identical in both cases. Even the concepts involved in cause of death in humans (e.g., Moriyama, 1956; Anon, 1962; Kitagawa, 1977) or differences in susceptibility to death (e.g., Vaupel and Yashin, 1985) have direct bearing on coding, classifying, and interpreting mortality patterns in insect populations. Distinguishing between the underlying cause of death and contributory causes of death is often as difficult in humans as it is in animals. In short, multiple decrement theory is as relevant to insects and populations of other non-human species as it is to humans.

SELECTED PROPERTIES OF MODEL LIFE TABLES

The life table is often thought of more as an organizational framework than as a model, but it is both. In this section I present three examples of the use of the life table in the context of a model: i) a statistical model; ii) sensitivity analysis, which relates the effect of a small change in mortality at a particular age to expectation of life; and iii) entropy, which characterizes the broad pattern of mortality over the entire life course.

Life Table Statistics

The pioneers of life table techniques were actuaries who had little need for application of probability theory or statistical methods. Actuaries typically use the weight of large numbers on which to base their arguments. On the other hand, biologists often use life tables as a bioassay tool where the number of individuals is less than 50. For these cases a statistical perspective for analysis of mortality is needed.

The life table is similar to statistical reliability theory in that life is a random experiment; its outcomes, survival and death, are subject to chance (Chiang, 1984). The period mortality correspond to failure rate. A statistical perspective on life table analysis is important for several reasons. *First*, demographic techniques given earlier provide specific methodologies for collecting, compiling, organizing, and analyzing demographic data in more of a deterministic context (e.g., l_x = fraction of a cohort surviving to age x). However, this fraction also represents a probability. *Second*, hypothesis testing is fundamental to science. A statistical perspective for life table analysis provides an epistemological link to other experimental and theoretical sciences. *Third*, many arithmetic and statistical techniques used in ecology and bioassay are related to those in demographic analysis. Therefore, the formal statistical bases of demography must be established before these ties can be fully understood. Two statistical perspectives on the life table will be introduced in this section—i) probability and ii) mean and variance. More detailed aspects of life table statistics can be obtained in Chiang (1984).

Probability
Probability is defined as the number of favorable outcomes in a series of experiments divided by the total number of possible outcomes. Consider an

Table 2-8. Summary of Two-Spotted Spider Mite Sex-Specific Survival and Mortality by Sex to Age 21 Days (Data from Hamilton, 1984)

Sex	Alive (A)	Dead (A')	Total
Female	13	12	25
(F)	n(FA)	n(FA')	n(F)
Male	42	11	53
(F')	n(F'A)	n(F'A')	n(F')
Total	55	23	78
	n(A)	n(A')	n

adult cohort of the two-spotted spider mite, *Tetranychus urticae*, consisting of 53 males and 25 females held at 27°C. After 21 days, it is observed that 12 of the females and 11 of the males have died (data from A. Hamilton, M. S. Thesis, UCD, 1984). These data are summarized in the Table 2-8, where F = female, F' = not female (= male), A = alive, and A' = not alive (= dead) and n = number observed in the specified category.

The probability of a mite's dying in this experiment, denoted Pr(A'), is equal to the number of dead mites at the end of 21 days divided by the total number subject to death over the period:

$$Pr(A') = n(A')/n$$
$$= 23/78$$
$$= .29$$

This is called *a posteriori probability* or frequency as distinct from classical, or *a priori probability*. The later type of probability is associated with coin tosses or dice rolling in basic statistical theory. For example, given a well-balanced coin, one would expect that the coin is just as likely to fall heads as tails; hence, the probability of the event of a head is given the value of $\frac{1}{2}$. There is no need to flip a coin several hundred times to determine this probability. Examples of classical probability used in biology include Hardy-Weinberg or Mendelian proportions in genetics and the normal distribution. Typically a test is conducted to tell whether an observed distribution departs from the classical, or "expected", distribution.

Mortality data is a form of a frequency or *a posteriori probability* since it is necessary to first observe death rates before we can make statements regarding how other groups of mites might react (survive) under similar conditions. The probability of a mite's being alive is

$$Pr(A) = n(A)/n$$
$$= 55/78 \qquad (2\text{-}13a)$$
$$= .71$$

or of a mite's being a female is

$$\begin{aligned} \Pr(F) &= n(F)/n \\ &= 25/78 \\ &= .32 \end{aligned} \qquad (2\text{-}13b)$$

The complements of two of these probabilities can be given as

$$\begin{aligned} \Pr(F) &= 1 - \Pr(F) \\ &= 1 - (25/78) \\ &= .68 \end{aligned} \qquad (2\text{-}13)c$$

$$\begin{aligned} \Pr(A') &= 1 - \Pr(A) \\ &= 1 - (55/78) \\ &= .29 \end{aligned} \qquad (2\text{-}13d)$$

The probability of not being a female (i.e., being a male) is 1 minus the probability of being a female, and the probability of not being alive (i.e., being dead) is 1 minus the probability of being alive.

Mean and variance

There are three different types of arithmetic means: i) expectation of a sample proportion; ii) expectation of a sample sum; and iii) expectation of a random variable. The distinction among these three types of means may best be explained by considering the following. Suppose the length of life of each individual adult fly in a large group was determined by an experimenter who rolled a fair die as each one eclosed. The number of days the individual was "allowed" to live would correspond to the number that turned up. Thus each fly would have an equal probability of living 1, 2, 3, 4, 5, or 6 days. The *expectation of a sample proportion* will be the proportion of flies expected to live for, say, 3 days. This is expressed as

$$E(p) = p \qquad (2\text{-}14)$$

and equals $\frac{1}{6}$ in this case. This probability is identical with all of the others since each number is equally likely to be rolled.

The *expectation of a sample sum* is equal to the number of times a particular number will appear out of n trials (flies). This is expressed as

$$E(x_1 + x_2 + \cdots + x_n) = np \qquad (2\text{-}15)$$

and equals 17 flies out of 100 trials for the current case (i.e., one-sixth of 100). This represents the expected number of times the number 3 appeared out of the 100 rolls of the die. This is the number of flies that would be allowed to live 3 days.

The expectation of a random variable equals the sum of the product of the probabilities (frequencies) and the values and is expressed as

$$E(x) = \sum_{k} k \Pr(x = k) \qquad (2\text{-}16)$$

where k is the number of possible outcomes. The value of k weights probability (i.e., proportion) by the number of days. In the current case for the die throws determining longevity gives an average life expectancy of

1 day $(k = 1) \times$ frequency of occurrence $(\Pr[x = 1]) = 1 \times 1/6 = .167$
2 days $(k = 2) \times$ frequency of occurrence $(\Pr[x = 2]) = 2 \times 1/6 = .333$
3 days $(k = 3) \times$ frequency of occurrence $(\Pr[x = 3]) = 3 \times 1/6 = .500$
4 days $(k = 4) \times$ frequency of occurrence $(\Pr[x = 4]) = 4 \times 1/6 = .667$
5 days $(k = 5) \times$ frequency of occurrence $(\Pr[x = 5]) = 5 \times 1/6 = .833$
6 days $(k = 6) \times$ frequency of occurrence $(\Pr[x = 6]) = 6 \times 1/6 = 1.000$

Thus

$$E(x) = \sum_k k \Pr[x = k]/k \qquad (k = 1, 2, \ldots, 6)$$
$$= .167 + .333 + .500 + .667 + .833 + 1.000$$
$$= 3.5 \text{ days}$$

Therefore the average fly lives 3.5 days.

In summary, the three averages in the demographic context are—

1. *Expectation of a sample proportion*—the *proportion* of all individuals that lived over the specified period.
2. *Expectation of a sample sum*—the *number* of individuals which are likely to live over the specified period.
3. *Expectation of a random variable*—the average number of individuals that experienced all events averaged.

The variance of each of these means is defined as

variance of a sample proportion $= \text{Var}(p) = pq/n$
variance of sample sum $= \text{Var}(x_1 + \cdots x_n) = npq$
variance of random variable $= \sigma^2 = \sum_k [k - E(x)]^2 \Pr(x = k)$

The standard deviation (SD) of each is the square root of their variance.

In order to illustrate each of these statistical measures, suppose that a group of 100 insects exhibited the life table characteristics shown in Table 2-9.

Table 2-9. Hypothetical Life Table for 100 Insects

x	l_x	d_x
0	100	26
1	74	18
2	56	14
3	42	24
4	18	18
5	0	0

LIFE TABLES

The expectation of sample proportion living to age class 3 and the variance is

$$E(2) = p = 56/100 = .56$$
$$Var(p) = np(1 - p)$$
$$= 100(.56)(.44)$$
$$= .0025$$
$$SD = \sqrt{.0025}$$
$$= .05$$

Thus the interrelation between the expectation of the sample proportion and the expectation of the sample sum in this example is as follows. Assuming that the hundred individuals in the hypothetical study are typical of all other groups of one hundred, 67% of the time the fraction of the cohort surviving to age class 3 should be within 1 SD of $p = 0.56$: i.e., between $p = 0.51$ (i.e., $p = .56 - .05$) and $p = .61$ (i.e., $p = .56 + .05$). Likewise, the number of individuals surviving to age 3 should be within 1 SD of the mean number out of 100 or between 51 and 61 individuals.

The random variable in a life table is death, and its probability distribution is the d_x schedule. Hence the mean age of death at birth is the life table parameter e_0. That is,

$$\text{mean age of death} = e_0$$
$$= \sum_{x=0}^{\omega} xd_x$$
$$= 0(.26) + 1(.18) + 2(.14) + 3(.24) + 4(.18)$$
$$= 1.9$$

Thus x in the equation is the age interval in which death may occur and d_x is the probability that death of a newborn will occur in the interval x to $x + 1$. The variance of deaths around this mean age is

$$\text{Variance} = \sum_{x=0}^{\omega} (x - e_0)^2 d_x$$
$$= (0 - 1.9)^2(.26) + (1 - 1.9)^2(.18) + (2 - 1.9)^2(.14)$$
$$+ (3 - 1.9)^2(.24) + (4 - 1.9)^2(.18)$$
$$\sigma^2 = 2.17$$

As an example problem, suppose the death rate of a group of insects is normally distributed with $\mu = 20$ days and $\sigma = 6$: i) What fraction of all deaths occur before 10 days? Before 35 days? ii) What fraction of deaths occur between ages 15 and 25 days?

The first question calls for the probabilities

$$\Pr(x < 10) \text{ and } \Pr(x < 35)$$

Converting x to standard normal random variables yields

$$\Pr(Z < [10 - 20]/6) \text{ and } \Pr(Z < [35 - 20]/6)$$

or
$$\Pr(Z < -1.67) \text{ and } \Pr(Z < 2.5)$$

Consulting a normal distribution table reveals that
$$\Pr(Z < -1.67) = .05$$
and
$$\Pr(Z < 2.50) = .995$$

Thus 5% of all deaths occur before age 10 days and 99.5% of all deaths are expected to have occurred by age 35 days.

The second question calls for the probability
$$\Pr(15 < x < 25)$$

Converting to the Z scale yields
$$\Pr(Z < [15-20]/6) \text{ and } \Pr(Z < [25-20]/6)$$
$$\Pr(-0.83 < Z < 0.83)$$

The normal distribution table shows that the area up to $Z = -.83$ is .197. Since the distribution is symmetrical, this will also equal the area greater than 0.83. Therefore
$$\Pr[-.83 < Z < .83] = 1.0 - 2(.197)$$
$$= .61$$

Thus 61% of all deaths would be expected to occur between the ages of 15 and 25 days.

Sensitivity Analysis

A question of importance in the analysis of survivorship involves the extent to which a slight change in survival at a specified age changes the expectation of life at birth, e_0. Assuming that the age interval is short, we can examine this in terms of l_x's.

$$e_0 = \sum_{x=0}^{\omega} l_x$$
$$= \sum_{x=0}^{\omega} l_0 \prod_{y=0}^{x-1} p_y$$
$$= 1 + p_0 + p_0 p_1 + p_0 p_1 p_2 + \cdots \qquad (2\text{-}14)$$

To determine the effect of a small change in, for example, p_1 of a four-age-class cohort, we set
$$e_0 = 1 + p_0 + p_0 p_1 + p_0 p_1 p_2$$
and
$$de_0/dp_1 = p_0 + p_0 p_2$$

LIFE TABLES

The derivative can also be expressed as

$$de_0/dp_1 = (1/p_1) \sum_{x=2}^{3} l_x$$

since

$$p_0 + p_0 p_2 = (p_0 p_1/p_1) + (p_0 p_1 p_2)/p_1$$

The term $(1/p_1)$ can be factored out of the right-hand side, yielding

$$(1/p_1)(p_0 p_1 + p_0 p_1 p_2)$$

or

$$(1/p_1)(l_2 + l_3)$$

Therefore the general form is

$$de_0/dp_x = (1/p_x) \sum_{y=x+1}^{\omega} l_y \qquad (2\text{-}15)$$

This expression illustrates two important aspects of the sensitivity of e_0 to changes in period survival. *First*, as x increases, the sum of the l_x's from x to ω continually decreases. Therefore the effect of a change in period survival on e_0 will always be greater at young ages than at older ages, all else being equal. *Second*, e_0 will be most greatly affected by changes in period survivorships (p_x's) that are low rather than those that are high. This is evident by noting that the term outside the summation is an inverse of a fraction. For example, if $p_x = .9$ the factor by which the sum of l_x's will be multiplied is 1.1 ($= 1/.9$). On the other hand if $p_x = .5$, the sum will be multiplied by 2.0 ($= 1/.5$). The reason for this is that a "small" change is a greater proportion of p_x when p_x is small than when it is large. In short, small changes in survival at young ages when mortality is high will most greatly affect the expectation of life at age 0.

Life Table Entropy

If all individuals die at exactly the same age, the shape of the l_x schedule is "rectangular," whereas if all individuals have exactly the same probability of dying at each age (i.e., all p_x's are identical), the l_x schedule decreases geometrically. The distribution of deaths by age varies greatly between the two patterns. A measure of this heterogeneity is referred to as entropy (H). Demitrius (1978, 1979) is recognized as originating this concept as applied to demographic heterogeneity. Goldman and Lord (1986) provided more intuitive interpretations of the measure. The formula for this measure using life table notation is

$$H = \left\{ \sum_{x=0}^{\omega} e_x d_x \right\} \Big/ e_0 \qquad (2\text{-}16)$$

The numerator (i.e., sum of products $e_x d_x$) can be interpreted in three different ways: i) the weighted average of life expectancies at age x; ii) the average days of future life that are lost by the observed deaths; or iii) the average number of days an individual could expect to live, given a second chance on life. The denominator is the expectation of life at birth, e_0, and thus converts the absolute effect to a relative effect.

Vaupel (1986) provided three different interpretations of entropy, H: i) the proportional increase in life expectancy at birth if every individual's first death were averted; ii) percentage change in life expectancy produced by a reduction of 1% in the force of mortality at all ages; and iii) the number of days lost owing to death per number of days lived. In general, entropy serves as a quantitative characterization of survival pattern. If $H = 0$, then all deaths occur at exactly the same age, and if $H = 1$, then the l_x schedule is exponentially declining. The intermediate value, $H = 0.5$, indicates a linear l_x schedule.

As an example, consider entropy for the human louse life table given in Table 2-2:

$$H = (e_0 d_0 + e_1 d_1 + \cdots + e_{64} d_{64} + e_{65} d_{65})/e_0$$
$$= [32.438(.014) + 31.891(.013) + \cdots + 1.000(.001) + .500(.001)]/32.438$$
$$= 11.794/32.438$$
$$= .364 (= \text{entropy})$$

The louse survival schedule and various entropy values are presented in Figure 2-3. Note that for reference when $H = 0$ all individuals die at once, thus heterogeneity in the death rate is nil. When $H = 1$ the number of days

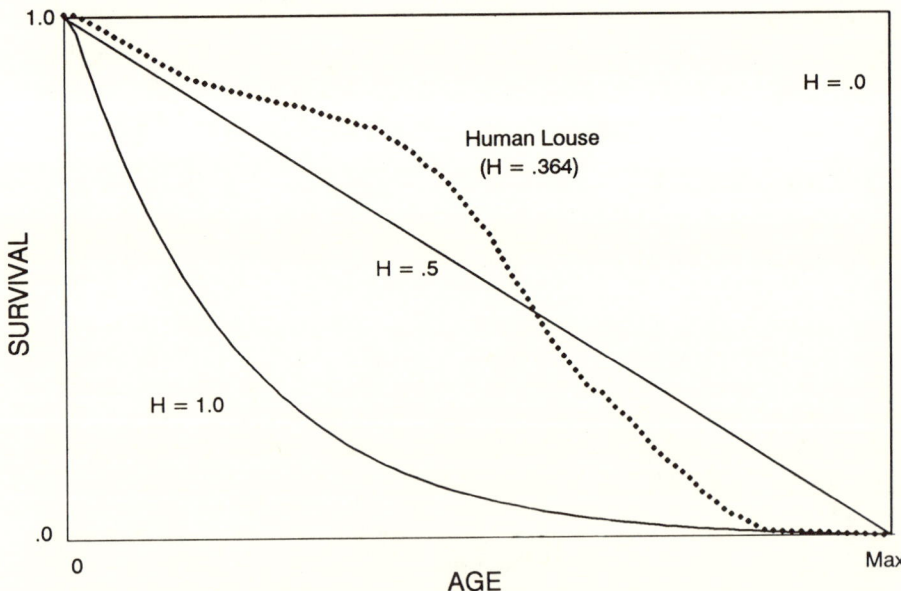

Figure 2-3. Shapes of three hypothetical life table survival schedules and associated entropy values, H. Life table of human louse taken from survival schedule given in Table 2-2.

lost in the cohort owing to death equals the average number of days lived by a newborn. The case of H = .5 is intermediate between the two extremes. The entropy value for the louse life table states that .36 days would be gained by the average individual if every first death were averted.

BIBLIOGRAPHY

Abbott, W. S. (1925) A method of computing the effectiveness of an insecticide. *J. Econ. Entomol.* 18:265–267.
Anon. (1962) *Vital Statistics Instruction Manual.* Part II. Procedures for Coding Multiple Causes of Deaths Occurring in 1955. National Office of Vital Statistics, Washington, D.C.
Batten, R. W. (1978) *Mortality Table Construction.* Prentice-Hall, Inc., Englewood Cliffs, New Jersey.
Brass, W. (1971) On the scale of mortality. In: *Biological Aspects of Demography.* W. Brass (ed.), pp. 69–110. Taylor and Francis, Ltd., London.
Cameron, P. J., and F. O. Morrison (1977) Analysis of mortality in the apple maggot, *Rhagoletis pomonella* (Diptera: Tephritidae), in Quebec. *Can. Entomol.* 109:769–788.
Carey, J. R. (1986) Interrelations and applications of mathematical demography to selected problems in fruit fly management. In: *Pest Control: Operations and Systems Analysis in Fruit Fly Management.* M. Mangel, J. Carey, and R. Plant (eds.), pp. 227–262. Springer-Verlag, Berlin.
Carey, J. R. (1989) The multiple decrement life table: A unifying framework for cause-of-death analysis in ecology. *Oecologia* 78:131–137.
Chiang, C. L. (1984) *The Life Table and its Applications.* Robert E. Krieger Publishing Co, Malabar, Florida.
DeBach, P., and C. B. Huffaker (1971) Experimental techniques for evaluation of the effectiveness of natural enemies. In: *Biological Control.* C. B. Huffaker (ed.), pp. 113–140. Plenum Press, New York.
Deevey, E. S., Jr. (1947) Life tables for natural populations of animals. *Q. Rev. Biol.* 22:283–314.
Demetrius, L. (1978) Adaptive value, entropy, and survivorship curves. *Nature* 257:213–214.
Demetrius, L. (1979) Relations between demographic parameters. *Demography* 16:329–338.
Elandt-Johnson, R. C. and N. L. Johnson (1980) *Survival Models and Data Analysis.* John Wiley and Sons, New York.
Evans, F. C. and F. E. Smith (1952) The intrinsic rate of natural increase for the human louse, *Pediculus humanus. Am. Natur.* 86:299–310.
Farr, W. (1875) Effect of the extinction of any single disease on the duration of life. *Suppl. 35th Ann. Rep. Registrar General* 21:38.
Finney, K. J. (1964) *Probit Analysis.* Cambridge University Press, Cambridge.
Goldman, N., and G. Lord (1986) A new look at entropy and the life table. *Demography* 23:275–282.
Hakkert, R. (1987) Life table transformations and inequality measures: Some noteworthy formal relationships. *Demography* 24:615–622.
Harcourt, D. G. (1969) The development and use of life tables in the study of natural insect populations. *Annu. Rev. Entomol.* 14:175–196.

Hewlett, P. S., and R. L. Plackett (1959) A unified theory for quantal responses to mixtures of drugs: Non-interactive action. *Biometrics* 15:591–610.

Huffaker, C. B., and C. E. Kennett (1966) Studies of two parasites of olive scale, *Parlatoria oleae* (Colvee). IV. Biological control of *Parlatoria olea* (Colvee) through the compensatory action of two introduced parasites. *Hilgardia* 37:283–335.

Keyfitz, N. (1982) Mathematical demography. In: *International Encyclopedia of Population*. J. A. Ross (ed.), pp. 437–443. Free Press, New York.

Keyfitz, N. (1985) *Applied Mathematical Demography*. 2nd ed. Springer-Verlag, Berlin.

Keyfitz, N., and J. Frauenthal (1975) An improved life table method. *Biometrics* 31:889–899.

Kitagawa, E. M. (1977) On mortality. *Demography* 14:381–389.

Luck, R. R., B. Shepard, and P. Kenmore (1988) Experimental methods for evaluating arthropod natural enemies. *Ann. Rev. Entomol.* 33:367–391.

Manton, K. G. and E. Stallard (1984) *Recent Trends in Mortality Analysis*. Academic Press, Inc., Orlando, Florida.

Moriyama, I. M. (1956) Development of the present concept of cause of death. *Am. J. Public Health* 46:436–441.

Morris, R. F. (1965) Contemporaneous mortality factors in population dynamics. *Can. Entomol.* 97:1173–1184.

Namboodiri, K., and C. M. Suchindran (1987) *Life Table Techniques and Their Applications*. Academic Press, Inc., Orlando, Florida.

Pollard, J. H. (1988) On the decomposition of changes in expectation of life and differentials in life expectancy. *Demography* 25:265–276.

Pressat, R. (1985) *The Dictionary of Demography*. Bell and Bain, Ltd., Glasgow.

Preston, S. H., N. Keyfitz, and R. Schoen (1972) *Cause of Death: Life Tables for National Populations*. Seminar Press, New York.

Royama, T. (1984) Population dynamics of the spruce budworm, *Choristoneura fumiferana*. *Ecol. Monographs* 54:429–462.

Sakagami, S. F. and H. Fukuda (1968) Life tables for worker honeybees. *Res. Popul. Ecol.* 10:127–139.

Schoen, R. (1975) Constructing increment-decrement life tables. *Demography* 12:313–324.

Southwood, R. R. E. (1971) *Ecological Methods*. 3rd ed. Chapman and Hall, London.

Thompson, W. R. (1955) Mortality factors acting in sequence. *Can. Entomol.* 87:264–275.

Varley, G. C. (1947) The natural control of population balance in the knapweed gall-fly. (*Urophora jacenana*). *J. Anim. Ecol.* 16:139–187.

Vaupel, J., and A. I. Yashin (1985) Heterogeneity's ruses: Some surprising effects of selection on population dynamics. *Am. Stat.* 39:1–25.

Vaupel, J. W. (1986) How change in age-specific mortality affects life expectancy. *Pop. Studies* 40:147–157.

3
Reproduction

Fertility rates were ignored historically probably because they had nothing directly to do with annuities. H. Nicander published a paper on fertility rates around 1800. These were used by Milne in 1815 in his treatise on annuities and by Böckh in 1886, who introduced what is now known as the net reproductive rate—expected progeny to an individual just born.

Reproduction can be characterized in two ways: i) the *process* that results in offspring; and ii) the per capita *rate* of offspring production in a given period of time. The first is physiological and the second is demographic. We are concerned here with the latter. Thus *reproductive rate*, *renewal rate*, *recruitment rate*, and *natality rate* all denote the same thing and are frequently used synonymously. Although *fecundity* and *fertility* are sometimes used interchangeably with reproductive rate, I define the former as the number of eggs laid in a specified age interval and the latter as the number of viable eggs laid in the interval. This distinction is useful to separate eggs laid from those that are laid but do not hatch. This distinction obviously does not apply to organisms that do not produce eggs.

Analysis of reproduction in this chapter is primarily descriptive since little modeling work has been done on reproduction in the entomological or ecological literature in a demographic context. Analytical approaches to studying and characterizing reproduction in humans falls into four broad categories, including general reproductive traits and processes (Kuczynski, 1931; Wicksell, 1931; Pollard, 1948; Perrin and Sheps, 1964; Coale and Trussell, 1974; Westoff, 1986), analysis of birth intervals (Feeney, 1983; Feeney and Ross, 1984; Potter, 1963), male reproduction (Myers, 1941; Karmel, 1947), and fertility by contraception modeling (Perrin and Sheps, 1964; Hoem, 1970; Sheps and Menken, 1973; Menken, 1975; Potter, 1975; Wood and Weinstein, 1988).

My objective in this chapter is to provide a general overview of the analysis of reproduction in biological populations that goes beyond the treatment in standard biology or ecology texts. I divide the subject into three general areas—reproductive averages, such as gross and net reproduction; reproductive heterogeneity, where differences among individual reproductive rates are characterized, and generalizations of reproduction, including clutch size and parity progression.

GENERAL BACKGROUND

The per capita age-specific fecundity and fertility rates over specified age intervals or over a cohort's lifetime is referred to as the *reproductive age schedule*. A *gross schedule* is one in which mortality is not taken into account and a *net schedule* weights reproduction by the proportion or the number in the cohort that survive to each age class.

Measures of gross and net fecundity and fertility give the mean number of offspring produced by the cohort but do not express any of the reproductive heterogeneity in that not all individuals produce the same number of offspring. Parameters that express various aspects of this heterogeneity include *reproductive interval*, which gives the daily frequency of offspring production among individuals; *daily parity*, which denotes the fraction of the cohort that produce specified numbers of offspring at a given age; *cumulative parity*, which expresses the fraction of living individuals within a cohort that have laid a specified sum total of offspring at given ages; and the *concentration of reproduction*, which gives the frequency distribution of lifetime reproduction ranked by individual. The offspring produced by insect females are often deposited in clutches and thus *clutch size* designates the offspring groupings and can be expressed as an age schedule. Most insects produce more than one *offspring type*, the most common of which is sex.

All of the reproductive relations discussed here refer to those pertaining to females and are thus *maternity functions*. Although very little effort has been directed towards reproductive rates of males, a number of *paternity functions* can be patterned after these maternity relations.

PER CAPITA REPRODUCTIVE RATES

Daily Rates—Basic Notation and Schedules

The most basic reproductive measure is gross maternity, defined as the average number of offspring produced by a female in the interval x to x + 1, designated M_x, where

$$M_x = \frac{\text{total number of offspring produced by female cohort between x and x + 1}}{\text{total number of females in cohort at midpoint of interval x to x + 1}}$$

The number of females alive at the midpoint of the age interval is simply

$$\text{midpoint number} = \frac{(\text{number alive at exact age x}) + (\text{number alive at exact age x + 1})}{2}$$

which implies that the individuals that died in the interval did so midway through the interval. This assumption is identical with that for determining the number of insect-days lived in the age interval for computing L_x in the life table. The parameter L_x is frequently used as a multiplier of M_x and

Table 3-1. Hypothetical Data on Age-Specific Reproductive Parameters

	Exact Age at Count			
	0	1	2	3
Data				
Number females alive at exact age	100	98	94	90
Estimate of number alive at midpoint of interval		99	96	92
Total offspring counted at exact age	0	1500	1200	1100
Parameters				
Age interval		0–1	1–2	2–3
Age index (x)		0	1	2
L_x		.99	.96	.92
M_x		15.15	12.50	11.96
$L_x M_x$		15.00	12.00	11.00

other reproductive parameters to yield net maternity, which is simply the average daily production of offspring by females aged x weighted by the probability of their attaining age x from age 0:

$$\text{net maternity} = L_x M_x \qquad (3\text{-}1)$$

Examples of the computations are given in Table 3–1.

Note that the number of offspring counted at the exact age represents the total number of produced in the age class interval x to x + 1. For example, the 1500 offspring counted at x = 1 corresponds to those produced by the 99 females (midpoint estimate) living in the age interval 0 to 1, which is indexed as x = 0. Thus it is necessary to distinguish between *exact age* and the designated *age class x*, which is the index for the interval between two exact ages.

It is often useful to distinguish betweeen female (daughters) and male (sons) offspring, and for this I use the notation m_x^f to designate the former and m_x^m for the latter. Thus

$$m_x^f = \frac{\text{total number of daughters produced by female cohort from x to x + 1}}{\text{total number of females at midpoint of interval x to x + 1}}$$

$$m_x^m = \frac{\text{total number of sons produced by female cohort from x to x + 1}}{\text{total number of females at midpoint of interval x to x + 1}}$$

and

$$M_x = m_x^f + m_x^m \qquad (3\text{-}2)$$

The lowercase notation, m_x, is commonly used in the biology literature to designate female offspring per female age x and will be used interchangeably with m_x^f.

The fraction of newborn daughters produced by a female age x, designated s_x, is given as

$$s_x = m_x^f / M_x \tag{3-3}$$

If the primary sex ratio is 1:1 over all ages as is the case for most diploid insects, then

$$m_x^f = m_x^m = M_x / 2 \tag{3-4}$$

The hatch rate at age x, denoted h_x, gives the fraction of all eggs produced by the cohort that are viable (i.e., hatch).

The functions m_x^f, m_x^m, and M_x represents the *gross fecundity schedules*, while the products $h_x m_x^f$, $h_x m_x^m$, and $h_x M_x$ represents the *gross fertility schedules*. The net schedules for each case (excluding s_x) incorporate the L_x term, which weights reproduction by the fraction of the female cohort that survive to age x. That is, the *net fecundity* and *net fertility schedules* are given by the products $L_x M_x$ and $L_x h_x M_x$, respectively. These concepts are readily extended to include net schedules for m_x^f and m_x^m.

An example data set for reproduction (excluding hatch rates) of 30 adult female spider mites recorded at 1-day intervals is given in Table 3-2. Computing the parameters M_x, m_x^f, m_x^m, and s_x is straightforward once three items are tabulated (see bottom of Table 3-2): i) total females living at each exact age; ii) total number of offspring by sex for all females produced from the preceding day to the current one; and iii) estimates of the interval midpoint cohort numbers. For example, a total of 18 females were alive on day 7 and a total of 158 offspring were counted—100 daughters and 58 sons. Since 25 females were alive on the previous day (day 6), an estimated total of $[(25 + 18)/2] = 21.5$ female-days were spent in the age interval 6 to 7 (x = 6). Therefore

$$m_6^f = 100 \text{ daughters}/21.5 \text{ females}$$
$$= 4.7 \text{ daughters/female}$$
$$m_6^m = 58 \text{ sons}/21.5 \text{ females}$$
$$= 2.7 \text{ sons/female}$$

and

$$M_6 = m_6^f + m_6^m$$
$$= 4.7 \text{ daughters} + 2.7 \text{ sons}$$
$$= 7.4 \text{ total offspring/female}$$

The fraction of the offspring that were daughters was

$$s_6 = m_6^f / M_6$$
$$= 4.7/7.4$$
$$= .64 \text{ (fraction daughters)}$$

Table 3-2. Age-Specific Production of Daughters and Sons in 30 Atlantic Spider Mites, *Tetranychus turkestani*, Females (Data from Krainacker, Unpublished)

Mite No.	Offspring sex	Exact Age (Days) at Count																Total
		0	1	2	3	4	5	6	7	8	9	10	11	12	13	14	15	
1	F	0	3	4	4	5	7	6	7	8	6	0	—	—	—	—	—	50
	M	0	1	2	2	2	1	3	2	1	3	0	—	—	—	—	—	17
2	F	0	2	2	0	1	0	1	1	2	2	2	2	—	—	—	—	15
	M	0	2	1	3	0	2	3	2	2	2	3	2	—	—	—	—	22
3	F	0	4	6	5	2	2	5	6	—	—	—	—	—	—	—	—	30
	M	0	2	3	2	1	0	3	1	—	—	—	—	—	—	—	—	12
4	F	0	11	10	8	8	9	4	0	—	—	—	—	—	—	—	—	50
	M	0	6	3	2	4	1	1	0	—	—	—	—	—	—	—	—	17
5	F	0	1	5	4	6	4	5	6	6	5	3	—	—	—	—	—	45
	M	0	4	3	3	2	3	4	3	3	3	2	—	—	—	—	—	30
6	F	0	0	3	5	4	3	4	1	—	—	—	—	—	—	—	—	20
	M	0	4	2	4	4	4	3	0	—	—	—	—	—	—	—	—	21
7	F	0	5	3	—	—	—	—	—	—	—	—	—	—	—	—	—	8
	M	0	3	0	—	—	—	—	—	—	—	—	—	—	—	—	—	3
8	F	0	7	7	7	7	10	9	9	9	8	3	4	2	5	2	2	91
	M	0	4	3	1	4	3	3	5	3	3	7	4	3	2	6	6	57
9	F	0	0	2	6	6	6	9	9	8	7	7	4	6	4	6	5	85
	M	0	0	2	1	1	2	1	2	0	2	2	3	1	2	3	3	25
10	F	0	2	3	0	—	—	—	—	—	—	—	—	—	—	—	—	5
	M	0	1	1	1	—	—	—	—	—	—	—	—	—	—	—	—	3
11	F	0	5	5	3	0	0	0	0	—	—	—	—	—	—	—	—	13
	M	0	4	3	1	0	0	0	—	—	—	—	—	—	—	—	—	8
12	F	0	0	0	0	0	—	—	—	—	—	—	—	—	—	—	—	0
	M	0	2	1	1	0	—	—	—	—	—	—	—	—	—	—	—	4
13	F	0	15	11	11	3	5	8	5	7	6	5	6	—	—	—	—	82
	M	0	5	3	4	7	8	6	3	2	5	5	1	—	—	—	—	49
14	F	0	4	3	5	3	5	0	—	—	—	—	—	—	—	—	—	20
	M	0	2	2	3	2	2	0	—	—	—	—	—	—	—	—	—	11

Table 3-2. (Contd.)

Mite No.	Offspring sex	\multicolumn{17}{c}{Exact Age (Days) at Count}

Mite No.	Offspring sex	0	1	2	3	4	5	6	7	8	9	10	11	12	13	14	15	Total
15	F	0	16	6	8	4	3	0	3	0	7	0	4	—	—	—	—	51
	M	0	3	5	8	7	8	12	8	12	5	13	1	—	—	—	1	82
16	F	0	9	9	10	5	6	5	6	9	6	4	3	2	4	1	1	80
	M	0	4	2	2	3	4	4	3	4	6	5	4	4	3	3	2	53
17	F	0	7	5	7	7	6	6	7	5	8	7	6	5	8	5	4	93
	M	0	3	4	3	2	2	5	3	2	2	4	2	3	2	3	4	44
18	F	0	0	1	3	2	1	1	0	0	0	0	0	—	—	—	—	8
	M	0	4	7	4	1	1	0	0	0	0	0	0	—	—	—	—	17
19	F	0	8	6	4	5	6	7	5	5	6	7	6	2	5	2	—	74
	M	0	3	2	2	2	3	3	4	3	2	2	1	4	1	2	—	34
20	F	0	0	0	0	0	—	—	—	—	—	—	—	—	—	—	—	0
	M	0	0	1	1	0	—	—	—	—	—	—	—	—	—	—	—	2
21	F	0	2	2	4	4	4	5	4	1	—	—	—	—	—	—	—	26
	M	0	0	3	3	3	2	2	2	1	—	—	—	—	—	—	—	16
22	F	0	4	5	4	4	6	7	4	5	4	3	1	—	—	—	—	47
	M	0	3	2	3	3	2	1	2	1	2	2	1	—	—	—	—	22
23	F	0	4	3	—	—	—	—	—	—	—	—	—	—	—	—	—	7
	M	0	3	4	—	—	—	—	—	—	—	—	—	—	—	—	—	7
24	F	0	5	0	0	5	4	5	6	4	2	4	3	1	1	3	1	44
	M	0	4	4	1	2	2	2	2	3	7	0	2	4	7	3	6	49
25	F	0	10	7	8	6	7	7	6	8	4	0	1	—	—	—	—	64
	M	0	6	5	4	4	2	3	4	4	1	1	0	—	—	—	—	34
26	F	0	0	4	5	3	1	3	3	3	4	4	2	0	1	1	1	35
	M	0	6	3	3	3	3	3	5	4	3	1	2	4	1	3	6	51
27	F	0	2	2	2	2	2	0	2	0	—	—	—	—	—	—	—	12
	M	0	0	1	2	2	2	0	1	1	—	—	—	—	—	—	—	7
28	F	0	0	0	1	1	2	3	0	0	1	0	0	—	—	—	—	8
	M	0	0	1	1	0	1	0	0	1	0	0	0	—	—	—	—	4

x	0	1	2	3	4	5	6	7	8	9	10	11	12	13	14	15	Total
29 F	0	0	4	5	7	8	9	6	2	—	—	—	—	—	—	—	41
29 M	0	0	7	7	4	2	5	4	1	—	—	—	—	—	—	—	30
30 F	0	0	8	6	8	8	8	6	8	7	7	5	4	2	2	0	79
30 M	0	0	6	3	6	3	7	5	5	3	3	3	3	5	4	1	59
Living females	30	30	30	28	27	25	25	18	18	17	17	15	8	8	8	7	
Midpoint number	30.0	30.0	29.0	27.5	26.0	25.0	21.5	18.0	17.5	17.0	16.0	11.5	8.0	8.0	7.5		
Offspring	231	208	200	172	182	186	158	138	132	104	72	47	53	48	42		1973
Daughters	138	125	128	109	116	112	100	87	83	54	46	20	30	21	14		1183
Sons	93	83	72	63	66	74	58	51	49	50	26	27	23	27	28		790
Offspring per female M_x	7.7 M_0	6.9 M_1	6.9 M_2	6.3 M_3	7.0 M_4	7.5 M_5	7.4 M_6	7.7 M_7	7.5 M_8	6.1 M_9	4.5 M_{10}	4.1 M_{11}	6.6 M_{12}	6.0 M_{13}	5.6 M_{14}		97.6
Daughters per female m_x^f	4.6 m_0^f	4.2 m_1^f	4.4 m_2^f	4.0 m_3^f	4.5 m_4^f	4.5 m_5^f	4.7 m_6^f	4.8 m_7^f	4.7 m_8^f	3.2 m_9^f	2.9 m_{10}^f	1.7 m_{11}^f	3.8 m_{12}^f	2.6 m_{13}^f	1.9 m_{14}^f		56.5
Sons per female m_x^m	3.1 m_0^m	2.8 m_1^m	2.5 m_2^m	2.3 m_3^m	2.5 m_4^m	3.0 m_5^m	2.7 m_6^m	2.8 m_7^m	2.8 m_8^m	2.9 m_9^m	1.6 m_{10}^m	2.3 m_{11}^m	2.9 m_{12}^m	3.4 m_{13}^m	3.7 m_{14}^m		41.3
Fraction Daughters s_x	.60 s_0	.61 s_1	.64 s_2	.63 s_3	.64 s_4	.61 s_5	.64 s_6	.62 s_7	.63 s_8	.52 s_9	.64 s_{10}	.41 s_{11}	.58 s_{12}	.43 s_{13}	.34 s_{14}		

Lifetime Rates and Mean Reproductive Ages

Maternity schedules may be summed over the lifetime of a cohort to yield a number of different reproductive rates or, in combination with age weightings with the survival schedule, used to determine mean reproductive ages in the cohort. These parameters will be presented using only total offspring (M_x) since extensions to the case of two sexes of offspring is straightforward.

A hypothetical yet important lifetime reproductive rate is the *gross fecundity rate*, given by the formula

$$\text{gross fecundity rate} = \sum_{x=\alpha}^{\beta} M_x \qquad (3\text{-}5a)$$

Since M_x is the average number of offspring produced by (living) females age x, then the sum of this schedule over all age classes gives the lifetime production of offspring by an average female that lives to the last day of possible life in the cohort. The counterpart of gross fecundity rate is *gross fertility rate*, given by the formula

$$\text{gross fertility rate} = \sum_{x=\alpha}^{\beta} h_x M_x \qquad (3\text{-}5b)$$

which gives the total number of viable eggs an hypothetical cohort of females will produce if all live to the last age of reproduction. The ratio of gross fertility to gross fecundity gives the gross hatch rate:

$$\text{gross hatch rate} = \frac{\text{gross fertility}}{\text{gross fecundity}}$$

$$= \frac{\sum_{x=\alpha}^{\beta} h_x M_x}{\sum_{x=\alpha}^{\beta} M_x} \qquad (3\text{-}5c)$$

which will always be less than or equal to unity. This rate is different than the average of h_x's over all age classes since it weights hatch by the number of eggs produced at each age.

Two net reproductive rates include

$$\text{net fecundity rate} = \sum_{x=\alpha}^{\beta} L_x M_x \qquad (3\text{-}5d)$$

and

$$\text{net fertility rate} = \sum_{x=\alpha}^{\beta} L_x h_x M_x \qquad (3\text{-}5e)$$

which give the average lifetime production of eggs and fertile eggs, respectively, for a newborn female. The net rates can also be computed as the dividend of the totals produced by the entire cohort divided by the initial number in the cohort. These two rates can be defined alternatively as the

REPRODUCTION

per generation reproductive contribution of newborn (or newly enclosed) females to the next generation. Since net rates give the per capita lifetime contribution, these rates divided by the number of days lived by the average female will yield the average daily production. Thus

$$\text{eggs/female/day} = \frac{\sum_{x=\alpha}^{\beta} L_x M_x}{\sum_{x=\varepsilon}^{\omega} L_x} \qquad (3\text{-}6a)$$

$$\text{fertile eggs/female/day} = \frac{\sum_{x=\alpha}^{\beta} L_x h_x M_x}{\sum_{x=\varepsilon}^{\omega} L_x} \qquad (3\text{-}6b)$$

Note that the denominator for each equals the expectation of life at eclosion.

The mean ages of reproduction for each of the schedules is computed by dividing the sum of each of the schedules weighted by age class x by the sum of the unweighted schedules:

$$\text{mean age gross fecundity} = \frac{\sum_{x=\alpha}^{\beta} x M_x}{\sum_{x=\alpha}^{\beta} M_x} \qquad (3\text{-}7a)$$

$$\text{mean age gross fertility} = \frac{\sum_{x=\alpha}^{\beta} x h_x M_x}{\sum_{x=\alpha}^{\beta} h_x M_x} \qquad (3\text{-}7b)$$

$$\text{mean age net fecundity} = \frac{\sum_{x=\alpha}^{\beta} x L_x M_x}{\sum_{x=\alpha}^{\beta} L_x M_x} \qquad (3\text{-}7c)$$

$$\text{mean age net fertility} = \frac{\sum_{x=\alpha}^{\beta} x h_x L_x M_x}{\sum_{x=\alpha}^{\beta} h_x L_x M_x} \qquad (3\text{-}7d)$$

$$\text{mean age hatch} = \frac{\sum_{x=\alpha}^{\beta} x h_x}{\sum_{x=\alpha}^{\beta} h_x} \qquad (3\text{-}7e)$$

The mean age of gross fecundity will always be greater than the mean ages for all other schedules owing to either female mortality, lack of egg hatch, or both.

An example of the application of these demographic techniques is given in Table 3-3 for data collected on the Mediterranean fruit fly. The basic protocol involves arranging each of the single parameters L_x, M_x, and h_x in columns according to age x, computing the appropriate products, including age weightings, and tabulating the sums of columns. The values for each of the parameters are thus

$$\text{gross fecundity} = \sum_{x=0}^{85} M_x = 809.57 \text{ eggs}$$

$$\text{gross fertility} = \sum_{x=0}^{85} h_x M_x = 434.95 \text{ fertile eggs}$$

$$\text{gross hatch} = \text{gross fertility/gross fecundity}$$
$$= 434.95/809.57$$
$$= .537 \text{ (fraction hatch)}$$

These parameters suggest that females that lived to the last day of possible life (i.e., 85 days) could be expected to have laid 809.57 total eggs and 434.95 fertile eggs for a hatch rate of 53.7%.

The net reproductive rates for this cohort are

$$\text{net fecundity} = \sum_{x=0}^{85} L_x M_x = 464.51 \text{ eggs}$$

$$\text{net fertility} = \sum_{x=0}^{85} L_x h_x M_x = 279.97 \text{ fertile eggs}$$

This analysis suggests that lack of hatch alone reduced potential reproduction by approximately 46%, female mortality alone reduced potential reproduction by approximately 43%, and the combination of lack of hatch and female mortality reduced potential reproduction by around 65%.

Expectation of life at age 0, e_0, is

$$e_0 = \sum_{x=0}^{85} L_x = 48.75$$

which gives the average number of female-days lived by the average female from age class 0 to death. Therefore, daily reproduction is given as

$$\text{eggs/female/day} = (\text{net fecundity}/e_0)$$
$$= 464.51/48.75$$
$$= 9.53 \text{ eggs/female/day}$$

$$\text{fertile eggs/female/day} = (\text{net fertility}/e_0)$$
$$= 279.97/48.75$$
$$= 5.74 \text{ fertile eggs/female/day}$$

Table 3-3. Reproductive Parameters of a Wild Strain of the Mediterranean Fruit Fly Collected from Makaha, Hawaii (Data from Harris and Carey, Unpublished)

Age (1)	Single Schedules			Composite Schedules				Age Weightings				
	L_x (2)	M_x (3)	h_x (4)	L_xM_x (5)	h_xM_x (6)	L_xh_x (7)	$L_xh_xM_x$ (8)	xM_x (9)	xh_x (10)	xL_xM_x (11)	xh_xM_x (12)	xL_xh_x (13)	$xL_xh_xM_x$ (14)
0	1.00	.00	.00	.00	.00	.00	.00	.00	.00	.00	.00	.00	.00
1	1.00	.00	.00	.00	.00	.00	.00	.00	.00	.00	.00	.00	.00
2	1.00	.00	.00	.00	.00	.00	.00	.00	.00	.00	.00	.00	.00
3	1.00	.84	.00	.84	.00	.00	.00	2.52	.00	2.52	.00	.00	.00
4	1.00	4.19	.79	4.19	3.31	.79	3.31	16.76	3.16	16.76	13.24	3.16	13.24
5	1.00	10.30	.85	10.30	8.76	.85	8.76	51.50	4.25	51.50	43.80	4.25	43.80
6	1.00	9.84	.73	9.84	7.18	.73	7.18	59.04	4.38	59.04	43.08	4.38	43.08
7	1.00	11.92	.76	11.92	9.06	.76	9.06	83.44	5.32	83.44	63.42	5.32	63.42
8	1.00	10.62	.92	10.62	9.77	.92	9.77	84.96	7.36	84.96	78.16	7.36	78.16
9	1.00	9.78	.90	9.78	8.80	.90	8.80	88.02	8.10	88.02	79.20	8.10	79.20
10	.97	10.14	.79	9.84	8.01	.77	7.77	101.40	7.90	98.40	80.10	7.66	77.70
11	.97	9.92	.76	9.62	7.54	.74	7.31	109.12	8.36	105.82	82.94	8.11	80.41
12	.95	10.14	.85	9.63	8.62	.81	8.19	121.68	10.20	115.56	103.44	9.69	98.28
13	.92	8.44	.83	7.76	7.01	.76	6.44	109.72	10.79	100.88	91.13	9.93	83.72
14	.89	10.36	.64	9.22	6.63	.57	5.90	145.04	8.96	129.08	92.82	7.97	82.60
15	.86	13.09	.74	11.26	9.69	.64	8.33	196.35	11.10	168.90	145.35	9.55	124.95
16	.86	11.34	.68	9.75	7.71	.58	6.63	181.44	10.88	156.00	123.36	9.36	106.08
17	.84	11.90	.76	10.00	9.04	.64	7.60	202.30	12.92	170.00	153.68	10.85	129.20
18	.78	11.62	.80	9.06	9.30	.62	7.25	209.16	14.40	163.08	167.40	11.23	130.50
19	.78	11.79	.74	9.20	8.72	.58	6.81	224.01	14.06	174.80	165.68	10.97	129.39
20	.78	11.90	.82	9.28	9.76	.64	7.61	238.00	16.40	185.60	195.20	12.79	152.20
21	.78	10.55	.67	8.23	7.07	.52	5.51	221.55	14.07	172.83	148.47	10.97	115.71
22	.76	13.39	.78	10.18	10.44	.59	7.94	294.58	17.66	223.96	229.68	13.04	174.68
23	.76	13.36	.82	10.15	10.96	.62	8.33	307.28	18.86	233.45	252.08	14.33	191.59
24	.76	11.00	.75	8.36	8.25	.57	6.27	264.00	18.00	200.64	198.00	13.68	150.48
25	.73	13.26	.82	9.68	10.87	.60	7.94	331.50	20.50	242.00	271.75	14.97	198.50

Table 3-3. (Contd.)

Age (1)	Single Schedules				Composite Schedules						Age Weightings			
	L_x (2)	M_x (3)	h_x (4)	L_xM_x (5)	h_xM_x (6)	L_xh_x (7)	$L_xh_xM_x$ (8)	xM_x (9)	xh_x (10)	xL_xM_x (11)	xh_xM_x (12)	xL_xh_x (13)	$xL_xh_xM_x$ (14)
26	.73	13.85	.57	10.11	7.89	.42	5.76	360.10	14.82	262.86	205.14	10.82	149.76
27	.73	11.26	.65	8.22	7.32	.47	5.34	304.02	17.55	221.94	197.64	12.81	144.18
28	.73	10.96	.75	8.00	8.22	.55	6.00	306.88	21.00	224.00	230.16	15.33	168.00
29	.73	14.11	.64	10.30	9.03	.47	6.59	409.19	18.56	298.70	261.87	13.55	191.11
30	.70	13.65	.61	9.55	8.33	.43	5.83	409.50	18.30	286.50	249.90	12.81	174.90
31	.70	11.88	.71	8.32	8.43	.50	5.90	368.28	22.01	257.92	261.33	15.41	182.90
32	.70	10.31	.54	7.22	5.57	.38	3.90	329.92	17.28	231.04	178.24	12.10	124.80
33	.70	9.96	.54	6.97	5.38	.38	3.76	328.68	17.82	230.01	177.54	12.47	124.08
34	.70	9.58	.47	6.71	4.50	.33	3.15	325.72	15.98	228.14	153.00	11.19	107.10
35	.68	11.32	.46	7.70	5.21	.31	3.54	396.20	16.10	269.50	182.35	10.95	123.90
36	.68	12.56	.49	8.54	6.15	.33	4.18	452.16	17.64	307.44	221.40	12.00	150.48
37	.65	10.13	.50	6.58	5.07	.33	3.29	374.81	18.50	243.46	187.59	12.03	121.73
38	.65	7.71	.36	5.01	2.78	.23	1.80	292.98	13.68	190.38	105.64	8.89	68.40
39	.59	9.50	.50	5.60	4.75	.29	2.80	370.50	19.50	218.40	185.25	11.50	109.20
40	.59	8.55	.35	5.04	2.99	.21	1.77	342.00	14.00	201.60	119.60	8.26	70.80
41	.59	9.14	.30	5.39	2.74	.18	1.62	374.74	12.30	220.99	112.34	7.26	66.42
42	.59	12.18	.31	7.19	3.78	.18	2.23	511.56	13.02	301.98	158.76	7.68	93.66
43	.59	9.77	.40	5.76	3.91	.24	2.31	420.11	17.20	247.68	168.15	10.15	99.33
44	.59	10.23	.36	6.04	3.68	.21	2.17	450.12	15.84	265.76	161.92	9.35	95.48
45	.54	9.30	.39	5.02	3.63	.21	1.96	418.50	17.55	225.90	163.35	9.48	88.20
46	.54	6.25	.48	3.38	3.00	.26	1.62	287.50	22.08	155.48	138.00	11.92	74.52
47	.51	11.79	.37	6.01	4.36	.19	2.22	554.13	17.39	282.47	204.92	8.87	104.34
48	.51	10.37	.42	5.29	4.36	.21	2.22	497.76	20.16	253.92	209.28	10.28	106.56
49	.51	12.16	.49	6.20	5.96	.25	3.04	595.84	24.01	303.80	292.04	12.25	148.96
50	.51	9.79	.26	4.99	2.55	.13	1.30	489.50	13.00	249.50	127.50	6.63	65.00
51	.51	10.74	.33	5.48	3.54	.17	1.81	547.74	16.83	279.48	180.54	8.58	92.31
52	.46	11.91	.48	5.48	5.72	.22	2.63	619.32	24.96	284.96	297.44	11.48	136.76
53	.43	14.06	.40	6.05	5.62	.17	2.42	745.18	21.20	320.65	297.86	9.12	128.26

54	.43	9.44	.36	4.06	3.40	.15	1.46	509.76	19.44	219.24	183.60	8.36	78.84
55	.43	10.69	.30	4.60	3.21	.13	1.38	587.95	16.50	253.00	176.55	7.10	75.90
56	.43	11.94	.35	5.13	4.18	.15	1.80	668.64	19.60	287.28	234.08	8.43	100.80
57	.43	10.00	.31	4.30	3.10	.13	1.33	570.00	17.67	245.10	176.70	7.60	75.81
58	.43	10.19	.27	4.38	2.75	.12	1.18	591.02	15.66	254.04	159.50	6.73	68.44
59	.43	10.06	.40	4.33	4.02	.17	1.73	593.54	23.60	255.47	237.18	10.15	102.07
60	.41	9.07	.49	3.72	4.44	.20	1.82	544.20	29.40	223.20	266.40	12.05	109.20
61	.41	9.67	.28	3.96	2.71	.11	1.11	589.87	17.08	241.56	165.31	7.00	67.71
62	.41	8.47	.27	3.47	2.29	.11	.94	525.14	16.74	215.14	141.98	6.86	58.28
63	.38	9.46	.38	3.59	3.59	.14	1.37	595.98	23.94	226.17	226.17	9.10	86.31
64	.35	6.85	.35	2.40	3.59	.12	.84	438.40	22.40	153.60	153.60	7.84	53.76
65	.35	10.46	.30	3.66	3.14	.11	1.10	679.90	19.50	237.90	204.10	6.83	71.50
66	.35	7.85	.36	2.75	2.83	.13	.99	518.10	23.76	181.50	186.78	8.32	65.34
67	.35	7.46	.42	2.61	3.13	.15	1.10	499.82	18.14	174.87	209.71	9.85	73.70
68	.32	5.58	.58	1.79	3.24	.19	1.04	379.44	39.44	121.72	220.32	12.62	70.72
69	.27	4.70	.36	1.27	1.69	.10	.46	324.30	24.84	87.63	116.61	6.71	31.74
70	.24	6.44	.57	1.55	3.67	.14	.88	450.80	39.90	108.50	256.90	9.58	61.60
71	.22	5.75	.41	1.27	2.36	0.9	.52	408.25	29.11	90.17	167.56	6.40	36.92
72	.22	8.95	.57	1.97	5.10	.13	1.12	644.40	41.04	141.84	367.20	9.03	80.64
73	.22	5.25	.33	1.16	1.73	0.7	.38	383.25	24.09	84.68	126.29	5.30	27.74
74	.19	8.00	.52	1.52	4.16	.10	.79	592.00	38.48	112.48	307.84	7.31	58.46
75	.19	5.86	.51	1.11	2.99	.10	.57	439.50	38.25	83.25	224.25	7.27	42.75
76	.19	6.14	.53	1.17	3.25	.10	.62	466.64	40.25	88.92	247.00	7.65	47.12
77	.16	3.50	.29	.56	1.02	0.5	.16	269.50	22.33	43.12	78.54	3.57	12.32
78	.11	5.25	.62	.58	3.26	0.7	.36	409.50	48.36	45.24	254.28	5.32	28.08
79	.08	7.33	.45	.59	3.30	0.4	.26	579.07	35.55	46.61	260.70	2.84	20.54
80	.05	8.50	.47	.43	4.00	0.2	.20	680.00	37.60	34.40	320.00	1.88	16.00
81	.05	5.00	.20	.25	1.00	0.1	.05	405.00	16.20	20.25	81.00	.81	4.05
82	.03	13.00	.38	.39	4.94	0.1	.15	1066.00	31.16	31.98	405.08	.93	12.30
83	.03	15.00	.27	.45	4.05	0.1	.12	1245.00	22.41	37.35	336.15	.67	9.96
84	.03	14.00	.50	.42	7.00	0.2	.21	1176.00	42.00	35.28	588.00	1.26	17.64
85	.03	7.00	.29	.21	2.03	0.1	.06	595.00	24.65	17.85	172.55	.74	5.10
	48.75	809.57	43.02	464.51	434.95	26.63	279.97	33952.28	1634.53	14491.04	15502.64	716.95	7329.37

The mean ages for the fecundity and fertility schedules are computed as

$$\text{mean age gross fecundity} = \frac{\sum_{x=0}^{85} xM_x}{\sum_{x=0}^{85} M_x}$$

$$= 33952.28/809.57$$

$$= 41.94 \text{ days}$$

$$\text{mean age gross fertility} = \frac{\sum_{x=0}^{85} xh_xM_x}{\sum_{x=0}^{85} h_xM_x}$$

$$= 15502.64/434.95$$

$$= 35.67 \text{ days}$$

$$\text{mean age net fecundity} = \frac{\sum_{x=0}^{85} xL_xM_x}{\sum_{x=0}^{85} L_xM_x}$$

$$= 14491.04/464.51$$

$$= 31.20 \text{ days}$$

$$\text{mean age net fertility} = \frac{\sum_{x=0}^{85} xL_xh_xM_x}{\sum_{x=0}^{85} L_xh_xM_x}$$

$$= 7329.37/279.97$$

$$= 26.18 \text{ days}$$

The interpretation of these parameters is that the average age of reproduction for the hypothetical cohort that lives to its last day of possible life (i.e., mean age gross fecundity) is 42 days, this mean age is reduced to slightly over 35 days by lack of egg hatch alone but by a full 10 days to around 31 days of age owing to mortality alone. The combination of lack of egg hatch and female mortality set the actual mean age of reproduction (viable offspring production) at 26.18 days.

Previous and Remaining Reproduction at Age x

The net and gross rates of fecundity given in the previous sections shed light on reproduction at two different points on the age scale of the average individual—age 0 and age ω. That is, the net rate represents the expected

REPRODUCTION

future reproduction of the average individual age class 0 and the gross rate represents the *past* reproduction of an individual that lived to the last possible age ($=\omega$). These basic relations are not restricted to the end points and can be extended to every age. The reproductive levels attained by the average female age x are

$$\text{per capita fecundity to age } x = \sum_{y=\alpha}^{x} M_y \qquad (3\text{-}7)$$

and the expected remaining fecundity is given by

$$\text{expected remaining fecundity at age } x = \frac{1}{L_x} \sum_{y=x+1}^{\omega} L_y M_y \qquad (3\text{-}8)$$

Therefore, the expected total fecundity for an average female age x is simply the sum of the previous reproduction and expected future reproduction: i.e.,

$$\text{expected total fecundity at age } x = \sum_{y=\alpha}^{x} M_y + \frac{1}{L_x} \sum_{y=x+1}^{\omega} L_y M_y \qquad (3\text{-}9)$$

These relations are given for the Mediterranean fruit fly data presented in the last section and are contained in Table 3-4. Note the following: i) the expected fecundity past age 0 (newly enclosed flies) is equal to the net fecundity rate. This is 464.5 eggs/female; and ii) at age 45 days the average female alive at age 45 days has laid a total of 455.64 eggs for a daily average of almost exactly 10 eggs/day. The average female at this age is expected to lay another 208 eggs before she dies for a lifetime total of 664 eggs.

REPRODUCTIVE HETEROGENEITY

All aspects of the analysis of reproduction in the previous sections were concerned with the average female. That is, every parameter contained M_x, which is an average. The purpose of this section is to present methods for examining the variation in reproduction among individuals within a cohort. Many of the ideas were taken from the paper by Carey et al. (1988).

Birth Intervals

Daily reproduction by individuals may be thought of as two types of responses to age: i) quantal in that offspring are either produced or not produced; and ii) graded, which refers to the number of offspring produced. This perspective sheds light on comparative aspects of reproduction among cohorts not evident from analysis of per capita daily reproduction, M_x. For example, consider daily reproduction of two individuals (i.e., Nos. 1 and 2) within each of three hypothetical cohorts (A–C) over a 2-day period (Table 3-5).

Note that the total reproduction for the 2-day period is identical for all individuals and the averages are the same for both days in all three cohorts.

Table 3-4. Past and Future Reproductive Parameters for the Mediterranean Fruit Fly (Data from Harris and Carey, Unpublished)

Age	Basic Parameters		Total Fecundity to Age x	Expected Fecundity past Age x	Expected Total Lifetime Fecundity at Age x
	L_x	M_x			
0	1.00	.00	.00	464.51	464.51
1	1.00	.00	.00	464.51	464.51
2	1.00	.00	.00	464.51	464.51
3	1.00	.84	.84	463.67	464.51
4	1.00	4.19	5.03	459.48	464.51
5	1.00	10.30	15.33	449.18	464.51
6	1.00	9.84	25.17	439.34	464.51
7	1.00	11.92	37.09	427.42	464.51
8	1.00	10.62	47.71	416.80	464.51
9	1.00	9.78	57.49	407.02	464.51
10	.97	10.14	67.63	409.46	477.09
11	.97	9.92	77.55	399.55	477.10
12	.95	10.14	87.69	397.82	485.51
13	.92	8.44	96.13	402.36	498.49
14	.89	10.36	106.49	405.56	512.05
15	.86	13.09	119.58	406.62	526.20
16	.86	11.34	130.92	395.28	526.20
17	.84	11.90	142.82	392.79	535.61
18	.78	11.62	154.44	411.38	565.82
19	.78	11.79	166.23	399.59	565.82
20	.78	11.90	178.13	387.69	565.82
21	.78	10.55	188.68	377.14	565.82
22	.76	13.39	202.07	373.67	575.74
23	.76	13.36	215.43	360.32	575.75
24	.76	11.00	226.43	349.32	575.75
25	.73	13.26	239.69	350.41	590.10
26	.73	13.85	253.54	336.56	590.10
27	.73	11.26	264.80	325.30	590.10
28	.73	10.96	275.76	314.34	590.10
29	.73	14.11	289.87	300.23	590.10
30	.70	13.65	303.52	299.46	602.98
31	.70	11.88	315.40	287.57	602.97
32	.70	10.31	325.71	277.26	602.97
33	.70	9.96	335.67	267.30	602.97
34	.70	9.58	345.25	257.71	602.96
35	.68	11.32	356.57	253.97	610.54
36	.68	12.56	369.13	241.41	610.54
37	.65	10.13	379.26	242.43	621.69
38	.65	7.71	386.97	234.72	621.69
39	.59	9.50	396.47	249.10	645.57
40	.59	8.55	405.02	240.56	645.58
41	.59	9.14	414.16	231.42	645.58
42	.59	12.18	426.34	219.24	645.58

REPRODUCTION

Table 3-4. (Contd.)

Age	Basic Parameters		Total Fecundity to Age x	Expected Fecundity past Age x	Expected Total Lifetime Fecundity at Age x
	L_x	M_x			
43	.59	9.77	436.11	209.47	645.58
44	.59	10.23	446.34	199.24	645.58
45	.54	9.30	455.64	208.39	664.03
46	.54	6.25	461.89	202.13	664.02
47	.51	11.79	473.68	202.24	675.92
48	.51	10.37	484.05	191.86	675.91
49	.51	12.16	496.21	179.71	675.92
50	.51	9.79	506.00	169.92	675.92
51	.51	10.74	516.74	159.18	675.92
52	.46	11.91	528.65	164.57	693.22
53	.43	14.06	542.71	161.98	704.69
54	.43	9.44	552.15	152.53	704.68
55	.43	10.69	562.84	141.84	704.68
56	.43	11.94	574.78	129.91	704.69
57	.43	10.00	584.78	119.91	704.69
58	.43	10.19	594.97	109.72	704.69
59	.43	10.06	605.03	99.65	704.68
60	.41	9.07	614.10	95.44	709.54
61	.41	9.67	623.77	85.78	709.55
62	.41	8.47	632.24	77.32	709.56
63	.38	9.46	641.70	73.97	715.67
64	.35	6.85	648.55	73.46	722.01
65	.35	10.46	659.01	63.00	722.01
66	.35	7.85	666.86	55.14	722.00
67	.35	7.46	674.32	47.69	722.01
68	.32	5.58	679.90	46.56	726.46
69	.27	4.70	684.60	50.48	735.08
70	.24	6.44	691.04	50.33	741.37
71	.22	5.75	696.79	49.14	745.93
72	.22	8.95	705.74	40.18	745.92
73	.22	5.25	710.99	34.91	745.90
74	.19	8.00	718.99	32.42	751.41
75	.19	5.86	724.85	26.58	751.43
76	.19	6.14	730.99	20.42	751.41
77	.16	3.50	734.49	20.75	755.24
78	.11	5.25	739.74	24.91	764.65
79	.08	7.33	747.07	26.88	773.94
80	.05	8.50	755.57	34.40	789.97
81	.05	5.00	760.57	29.40	789.97
82	.03	13.00	773.57	36.00	809.57
83	.03	15.00	788.57	21.00	809.57
84	.03	14.00	802.57	7.00	809.57
85	.03	7.00	809.57	.00	809.57

Table 3-5. Example of Reproductive Heterogeneity in Three Hypothetical Cohorts of Insects

	Cohort A			Cohort B			Cohort C		
Day	1	2	Average	1	2	Average	1	2	Average
1	10	0	5	2	8	5	5	5	5
2	0	10	5	8	2	5	5	5	5
	10	10	10	10	10	10	10	10	10

Therefore neither the averages nor total reproduction shed light on the true reproductive differences among the three cohorts. Individuals in Cohort A produced offspring every other day, while those in Cohorts B and C produced offspring every day. In terms of similarities in birth interval, Cohorts B and C are more alike, but in terms of daily production of offspring Cohorts A and B are most similar in that the daily variation was the highest in these two.

In the analysis of birth intervals it is important to distinguish i) the period between the age of adulthood and the age of reproductive maturity in which no eggs are laid; and ii) the intermittent days past reproductive maturity of an individual in which it produces no offspring. The first period is usually thought of as a separate demographic parameter—the age of first reproduction, α. Therefore lumping the two types of zero production days confounds the analysis. The complete analysis of birth interval in a cohort requires the following steps:

Step 1. Determine age of first reproduction for each individual.
Step 2. Count the number of days each individual lived in the observation period.
Step 3. Compute the number of mature-days by subtracting the age of first reproduction from the total days lived.
Step 4. Count the number of days in which at least 1 offspring was produced.
Step 5. Compute the fraction (or percentage) of mature-days that at least 1 offspring was produced. The inverse of this value yields birth interval for each individual.

Computations for cohort birth interval for the medfly data in Table 3-6 are presented in Table 3-7. Tallies for each fly include the age of first egg (age of reproductive maturity), total days lived out of the 30 possible days, mature days lived, and the number of days in which at least 1 egg was laid. For example, female 1 produced her first egg while in age class 3, lived the full 30 days, and thus lived a total of 27 ($= 30 - 3$) mature-days. Of these mature days, on all but 2 days (age classes 20 and 29) at least one egg was laid, for a total of 25 laying days. Thus she spent 93% of all mature days laying ($= 100 \times 25/27$) for an individual birth interval of 1.08 days ($= 1/.93$).

The number of female-days spent in the cohort was 1,318, with 207 of the days spent in the prereproductive period. Thus the average individual lived 27.5 days out of the 30 possible days ($= 1318/48$). Those that reached

Table 3-6. Number of Eggs per Day Laid by 48 Individual Mediterranean Fruit Flies Observed for a Total of 30 Days

Exact Age (Days)

Fly Number	0	1	2	3	4	5	6	7	8	9	10	11	12	13	14	15	16	17	18	19	20	21	22	23	24	25	26	27	28	29	Total
1	0	0	0	13	24	41	88	43	67	47	30	39	56	27	37	43	25	48	30	40	0	22	35	25	50	37	28	50	27	0	972
2	0	0	0	10	36	68	29	34	35	30	36	40	36	26	34	30	15	30	20	44	41	28	24	40	23	32	8	38	35	33	855
3	0	0	0	5	67	26	45	29	33	40	45	57	15	34	33	37	23	49	25	33	44	36	22	31	20	27	11	25	19	37	838
4	0	0	0	0	30	44	50	56	32	10	25	16	37	36	38	60	37	47	26	27	14	34	19	28	19	16	17	30	28	27	833
5	0	0	0	0	27	25	28	50	42	7	21	11	46	38	20	36	49	32	20	34	39	37	40	20	25	0	35	45	38	51	816
6	0	0	0	0	45	44	37	73	20	0	15	20	47	41	18	39	27	24	39	38	26	26	18	16	9	28	35	27	36	40	801
7	0	0	0	0	27	49	57	27	13	36	41	42	55	28	14	42	26	33	16	34	40	19	17	21	13	45	8	30	17	37	796
8	0	0	0	13	0	17	53	72	38	31	35	38	41	46	29	27	34	37	20	28	20	34	14	13	22	31	28	25	20	17	789
9	0	0	0	0	24	29	19	30	17	10	13	8	33	55	31	31	34	24	6	47	48	32	29	34	18	30	35	15	45	30	781
10	0	0	0	0	43	31	40	45	19	31	40	35	20	21	27	10	28	34	47	37	24	30	18	18	27	24	8	16	24	29	745
11	0	0	0	0	8	13	29	27	47	25	38	37	30	37	45	24	27	11	40	38	25	17	28	26	39	16	40	15	36	21	740
12	0	0	0	0	37	61	28	58	30	32	31	24	40	19	20	23	20	39	37	43	30	14	15	15	17	27	15	16	—	—	688
13	0	0	0	0	27	26	34	45	19	20	46	16	36	17	16	33	26	27	36	18	31	30	16	24	16	20	29	20	16	17	661
14	0	0	0	0	0	44	32	25	27	21	40	32	20	13	59	17	29	10	11	6	4	24	28	30	35	28	27	30	20	37	649
15	0	0	0	0	0	45	0	45	28	29	28	36	5	15	30	19	35	38	30	19	29	26	21	19	31	33	27	31	10	30	638
16	0	0	14	0	0	63	0	40	25	26	33	13	70	21	18	24	0	15	18	0	26	21	22	27	27	38	21	19	20	21	633
17	0	0	13	0	52	24	23	27	35	24	35	32	41	20	25	27	14	17	11	28	14	5	14	25	16	32	30	22	—	—	629
18	0	0	0	0	40	27	25	24	25	35	11	10	30	16	24	21	31	33	28	21	0	23	20	16	36	34	30	35	20	19	622
19	0	0	0	0	0	0	19	38	37	39	36	34	20	47	14	13	16	16	13	0	15	18	15	36	38	38	18	55	7	13	614
20	0	0	0	0	14	60	18	29	18	29	27	20	36	38	30	30	14	10	27	15	28	10	17	31	29	30	22	13	19	27	611
21	0	0	0	0	25	8	37	30	10	27	26	38	32	30	15	12	7	18	44	30	23	20	17	27	29	17	20	20	19	37	610
22	0	0	0	0	0	66	16	30	18	20	23	14	25	15	19	14	39	6	28	25	20	39	27	29	14	14	16	36	26	34	605
23	0	0	0	0	74	17	20	14	5	15	41	37	39	35	25	6	20	8	8	19	28	40	9	17	14	0	20	38	0	29	603
24	0	0	0	0	23	35	40	38	17	14	30	40	38	27	50	63	21	40	17	19	30	28	9	7	15	—	29	—	—	—	601
25	0	0	0	20	33	43	60	28	32	44	16	0	0	19	4	14	14	6	0	14	47	35	14	28	27	22	26	15	11	27	596
26	0	0	0	7	28	28	37	17	22	6	12	21	0	32	19	34	18	15	35	29	27	18	29	29	29	24	23	30	31	29	586
27	0	0	0	0	18	19	25	35	38	76	19	25	39	14	20	18	15	19	35	28	12	29	37	37	13	0	0	25	19	—	564
28	0	0	0	0	0	0	10	28	36	22	27	16	10	28	30	15	0	30	39	14	37	20	43	20	32	32	10	21	15	39	563
29	0	0	0	0	0	0	35	27	25	22	14	10	19	15	15	13	11	30	36	34	22	20	18	12	20	15	26	30	17	15	557
30	0	0	0	9	36	40	36	18	23	24	37	19	47	16	12	22	10	25	17	35	26	18	14	23	12	20	16	21	30	27	537
31	0	0	0	0	22	49	19	25	30	40	44	30	23	47	30	17	42	11	0	26	28	23	11	11	8	10	13	19	0	—	525
32	0	0	0	0	0	37	36	46	40	34	16	24	23	26	18	26	22	30	48	0	28	0	23	0	10	11	0	17	0	0	520
33	0	0	0	0	0	20	18	54	46	25	38	14	27	18	11	25	30	29	12	15	0	9	14	0	11	0	24	0	11	30	512
34	0	0	0	5	28	18	47	15	16	0	9	17	26	12	29	9	15	24	18	24	40	9	40	10	8	11	24	10	26	35	509
35	0	0	0	0	16	16	20	28	34	32	24	18	29	22	17	34	24	9	27	40	17	15	—	30	18	15	17	2	17	—	492
36	0	0	0	12	14	14	20	30	14	24	37	28	26	32	0	48	0	43	14	10	27	—	—	18	—	18	—	—	—	—	460

Table 3-6. (Contd.)

Fly Number	Exact Age (Days)																														Total
	0	1	2	3	4	5	6	7	8	9	10	11	12	13	14	15	16	17	18	19	20	21	22	23	24	25	26	27	28	29	
37	0	0	0	0	0	34	49	0	16	0	17	30	21	19	18	14	5	40	28	35	29	28	17	14	0	19	0	0	0	—	433
38	0	0	0	0	0	59	10	24	10	17	36	35	16	25	35	25	0	29	40	17	0	0	16	0	20	0	0	0	0	—	414
39	0	0	0	0	10	20	22	18	0	15	0	23	0	0	0	0	26	28	30	40	20	18	30	20	20	30	0	0	8	0	370
40	0	0	0	0	0	15	18	0	0	0	28	29	18	13	21	11	10	20	8	21	30	17	28	7	18	0	0	0	0	0	320
41	0	0	0	0	22	70	21	36	10	30	15	23	5	17	10	18	15	16	15	4	0	8	0	0	0	0	0	0	0	—	259
42	0	0	0	0	0	15	30	17	16	21	15	15	16	16	11	0	12	0	14	26	0	26	25	0	0	0	0	0	0	—	248
43	0	0	0	0	0	55	0	9	8	0	23	0	0	9	4	0	5	0	0	0	0	0	0	0	0	0	0	0	10	0	202
44	0	0	0	0	13	8	20	25	0	0	12	0	10	0	0	0	0	0	0	0	0	0	0	0	0	0	0	0	0	27	139
45	0	0	0	0	0	0	32	0	0	0	0	5	0	0	5	0	0	0	0	0	0	—	—	—	—	—	—	—	—	—	37
46	0	0	0	0	4	0	0	—	—	—	—	—	—	—	—	—	—	—	—	—	—	—	—	—	—	—	—	—	—	—	5
47	0	0	0	0	0	0	—	—	—	—	—	—	—	—	—	—	—	—	—	—	—	—	—	—	—	—	—	—	—	—	4
48	0	0	0	0	0	—	—	—	—	—	—	—	—	—	—	—	—	—	—	—	—	—	—	—	—	—	—	—	—	—	0
Total eggs	0	0	0	107	839	1458	1332	1409	1073	1030	1170	1071	1203	1056	981	1045	854	1052	1013	1093	956	931	816	843	898	767	747	802	621	815	25982
Total alive females	48	48	48	48	48	48	46	46	46	46	46	46	45	45	45	44	44	44	44	44	44	42	42	42	42	42	40	36	35	34	1318
Midpoint number	48	48	48	48	48	48	47	46	46	46	46	45.5	45	45	44.5	44	44	44	44	44	43	42	42	42	42	41	38	35.5	34.5		1237.0
Eggs per female	0.0	0.0	2.2	17.5	30.4	27.8	30.0	23.7	22.4	25.4	23.3	26.4	23.5	21.8	23.5	19.4	23.9	23.0	24.8	21.7	21.7	19.4	20.1	21.4	18.3	18.2	21.1	17.5	23.6		591.6
M_x	M_0	M_1	M_2	M_3	M_4	M_5	M_6	M_7	M_8	M_9	M_{10}	M_{11}	M_{12}	M_{13}	M_{14}	M_{15}	M_{16}	M_{17}	M_{18}	M_{19}	M_{20}	M_{21}	M_{22}	M_{23}	M_{24}	M_{25}	M_{26}	M_{27}	M_{28}		

Table 3-7. Computation of Birth Interval for the Mediterranean Fruit Fly (Data from Table 3-6)

Individual	Age First Egg	Total Days Lived	Mature Days Lived	Number Days Laid	Percent Mature Days Laying
1	3	30	27	25	93
2	3	30	27	27	100
3	3	30	27	27	100
4	4	30	26	26	100
5	4	30	26	25	96
6	4	30	26	24	92
7	4	30	26	26	100
8	5	30	25	25	100
9	3	30	27	27	100
10	4	30	26	26	100
11	4	30	26	26	100
12	4	27	23	23	100
13	4	30	26	26	100
14	5	30	25	25	100
15	5	30	25	23	92
16	3	30	27	23	85
17	3	27	24	24	100
18	4	30	27	27	100
19	6	30	24	23	96
20	4	30	26	26	100
21	4	30	26	26	100
22	5	30	25	24	96
23	4	30	26	24	92
24	4	27	23	21	91
25	3	30	27	24	89
26	3	30	27	25	93
27	4	27	20	18	90
28	6	30	24	23	96
29	6	30	23	23	100
30	4	30	26	25	96
31	4	28	24	22	92
32	4	30	26	17	65
33	5	30	25	24	96
34	4	30	26	25	96
35	4	21	17	17	100
36	3	30	27	22	81
37	5	29	24	18	75
38	5	26	21	16	76
39	4	30	26	16	62
40	5	30	25	18	72
41	4	15	11	11	100
42	5	26	21	16	76
43	5	30	25	10	40
44	4	30	26	10	38
45	6	12	6	2	33
46	14	21	7	1	14
47	4	6	2	1	50
48		6	0	0	0
Total	207	1318	1102	983	

reproductive maturity before dying (fly 48 did not) spent an average of 4.4 days in the prereproductive period (= 207/47), 23.5 days in the reproductive period (= 1102/47), and 20.9 days reproducing.

The percentage of days in which reproductively mature females produced at least one egg was 89.2% (= 100 × 983/1102) for a birth interval of 1.12 days (= 1/.892). That is, on average a reproductively mature female produced some eggs every 1.12 days. Note that 18 reproductively mature individuals or 38% of the cohort produced eggs 100% of the time (i.e., every day) and 6 individuals or 13% of the cohort produced eggs less than half the time (i.e., females 43 to 48).

Reproductive Parity

Consider the following reproduction in an hypothetical cohort of four individuals over a 4-day period (Table 3-8).

All four individuals in this hypothetical cohort produce offspring every day; thus no differences exist among their birth intervals. Furthermore, average daily reproduction, M_x, is the same over all age classes. Yet it is clear that differences in reproduction do exist among the cohort members in that i) individual 1 consistently produced only one offspring/day and individual 4 consistently produced two offspring/day; and ii) individuals 2 and 3 were inconsistent on a daily basis but were consistent when their laying is considered over the entire 4-day period.

Differences in daily offspring production of individuals within a cohort can be expressed in terms of daily parity classes, which give the fraction of the cohort whose offspring production falls into one of several preselected reproductive classes. For example, the fraction of the hypothetical cohort producing either one or two offspring in any given age class is given in Table 3-9.

Thus at all ages exactly 50% of the cohort produced one offspring and 50% of the cohort two offspring, which shows why the average daily production, M_x, equals 1.5 for all age classes.

A second parity measures involves the cumulative daily reproduction of individuals and is thus termed cumulative parity. These measures express the fraction of the cohort whose previous and current cumulative offspring

Table 3-8. Example of Reproductive Heterogeneity in Four Hypothetical Female Insects

Age (Days)	Individual				Average
	1	2	3	4	
0	1	1	2	2	1.5
1	1	2	1	2	1.5
2	1	1	2	2	1.5
3	1	2	1	2	1.5
Total	4	6	6	8	6.0

Table 3-9. Summary of Daily Reproductive Parity for the Hypothetical Example Given in Table 3-8

	Cohort Producing—	
Age (Days)	1 Offspring	2 Offspring
0	50	50
1	50	50
2	50	50
3	50	50

Table 3-10. Summary of Cumulative Reproductive Parity for the Hypothetical Example Given in Table 3-8

	Cumulative Prodcution by Individual				Percentage of Cohort in Cumulative Class			
Age (Days)	1	2	3	4	1–2	3–4	5–6	>6
0	1	1	2	2	100	0	0	0
1	2	3	3	4	25	75	0	0
2	3	4	5	6	0	50	50	0
3	4	6	6	8	0	25	50	25

production falls into one of several preselected reproductive classes. Table 3-10 shows these relationships for the hypothetical data given above.

The value of the cumulative parity relations is most evident for age class 3, where individual 1 shows up as the quarter (25% in class 3–4) of the cohort whose offspring production is consistently low and individual 4 as the quarter (25% in class >6) that is consistently high.

Daily parity provides insight into the consistency of egg-laying levels in the cohort at a specific age, while cumulative parity shows the longer-term consistency. For example, the daily parity of a cohort in which each individual lays one egg every day will be different than a cohort in which the individuals lay exactly ten eggs on each day of egg laying but lay on average every tenth day. In both cases the mean eggs/female/day in the cohorts are identical ($=1$), as will be their cumulative parity. However, the differences in their laying strategies are obscured.

Choice of the number of daily parity classes and their divisions is judgmental, but at the minimum should include a zero class and two or three divisions. The medfly data from Table 3-6 was examined using four daily parity classes—0, 1–25, 26–50, and >50 eggs. The results are given in Table 3-11. For example, note that 9 individuals in age class 3 laid from 1 to 25 eggs (Nos. 1, 2, 3, 9, 16, 17, 25, 26, 36) and the remainder laid no eggs at this age. Since all 48 flies were alive, the percentage in each of the 4 daily parity classes was 81.2% ($=100 \times 39/48$) for egg class 0, 18.8% ($=100 \times 9/48$) for egg class 1–25, and 0% for egg classes 26–50 and >50.

Table 3-11. Daily Parity of the Mediterranean Fruit Fly (Data from Table 3-4)[1]

Age Class	Number of Females					Percentage of Females				
	Daily Egg Class					Daily Egg Class				
	0	1–25	26–50	>50	Total	0	1–25	26–50	>50	Total
0	48	0	0	0	48	100.0	0.0	0.0	0.0	100.0
1	48	0	0	0	48	100.0	0.0	0.0	0.0	100.0
2	48	0	0	0	48	100.0	0.0	0.0	0.0	100.0
3	39	9	0	0	48	81.2	18.8	0.0	0.0	100.0
4	18	15	12	3	48	37.5	31.2	25.0	6.3	100.0
5	7	14	19	8	48	14.6	29.2	39.5	16.7	100.0
6	4	18	19	5	46	8.7	39.1	41.3	10.9	100.0
7	4	12	24	6	46	8.7	26.1	52.2	13.0	100.0
8	5	22	18	1	46	10.9	47.8	39.1	2.2	100.0
9	8	19	18	1	46	17.4	41.3	39.1	2.2	100.0
10	4	17	25	0	46	8.7	37.0	54.3	0.0	100.0
11	4	23	18	1	46	8.7	50.0	39.1	2.2	100.0
12	5	16	21	3	45	11.1	35.5	46.7	6.7	100.0
13	3	23	18	1	45	6.7	51.1	40.0	2.2	100.0
14	3	26	14	2	45	6.7	57.8	31.1	4.4	100.0
15	4	23	15	2	44	9.1	52.3	34.1	4.5	100.0
16	6	22	16	0	44	13.6	50.0	35.4	0.0	100.0
17	3	20	21	0	44	6.8	45.5	47.7	0.0	100.0
18	4	19	21	0	44	9.1	43.2	47.7	0.0	100.0
19	4	15	25	0	44	9.1	34.1	56.8	0.0	100.0
20	8	14	22	0	44	18.2	31.8	50.0	0.0	100.0
21	3	21	18	0	42	7.1	50.0	42.9	0.0	100.0
22	5	24	13	0	42	11.9	57.1	31.0	0.0	100.0
23	5	22	15	0	42	11.9	52.4	35.7	0.0	100.0
24	5	22	14	1	42	11.9	52.4	33.3	2.4	100.0
25	10	17	15	0	42	23.8	40.5	35.7	0.0	100.0
26	8	16	15	1	40	20.0	40.0	37.5	2.5	100.0
27	6	15	13	2	36	16.7	41.6	36.1	5.6	100.0
28	7	17	11	0	35	20.0	48.6	31.4	0.0	100.0
29	6	7	20	1	34	17.6	20.7	58.8	2.9	100.0
Total	332	488	460	38	1318					

[1] Daily egg class refers to daily egg production of 0, 1–25, 26–50, or >50 eggs by day.

The analysis provides perspectives on the within cohort variation in egg-laying levels. For example, per capita egg production levels in age classes 4 (M_4) and 28 (M_{28}) from Table 3-6 were both 17.5 eggs/female but for different reasons. Nearly 40% of the individuals in age class 4 laid zero eggs, but only 20% of those in age class 28 laid zero eggs. The high percentage of individuals that laid zero eggs in age class 4 was offset by a substantial percentage that laid 26–50 eggs (25%) and a small percentage that laid over 50 eggs (6.3%).

These summary data also show a total of 1318 female-days in the cohort, of which 332, 488, 460, and 38 days were spent laying 0, 1–25, 26–50, and >50 eggs, respectively. Therefore over the entire period on about 25% of all days ($= 332/1318$) no eggs were laid, on 37% of the days 1–25 eggs were laid, on 35% of all days 26–50 eggs were laid, and on 3% of all days more than 50 eggs were laid. The number of days on which one or more eggs were laid was 986 ($= 1318 - 332$). Thus the distribution of egg production levels for those days in which at least one egg was laid was 49.5% for 1–25 eggs, 46.7% for 26–50 eggs, and 3.9% for >50 eggs.

Cumulative parity for the medly cohort is shown in Table 3-12. A total of six cumulative egg classes were chosen: 0, 1–200, 201–400, 401–600, 601–800, and >800. The spread in egg production rates among individuals can be seen by noting that 12.5% (6 females) of the cohort in age class 5 had laid no eggs, while the remaining 87% (42 individuals) of the cohort had laid between 1 and 200 eggs. None had laid over 200 eggs by this age. By age class 14 all females in the cohort that were still alive (i.e., 45 females) had laid some eggs. Of these, 20% (i.e., 9 females) had laid 1–200 eggs, 71% (i.e., 32 females) had laid 201–400 eggs, and 8.9% (i.e., 4 females) had laid 401–600 eggs. The spread in cumulative egg production becomes even greater by age class 24, when one individual (female 43) had still only produced a cumulative total of less than 200 eggs, while one other individual (female 1) had produced well over 800 total eggs. By the end of the 30-day period a total of 6 individuals (females 1–6) had laid over 800 eggs, while a total of 4 living individuals had laid 400 eggs or less (females 39, 40, 43 and 44).

GENERALIZATIONS

Reproduction is typically thought of in terms of offspring production by females of a given age. However, the application of methods presented in this chapter as well as several given in the previous chapter on life tables are not limited to the chief purpose for which they were developed. If g_x denotes any event at age x, then Table 3-13 gives the general formulae for determining event intensity and event timing.

Clutch

While analysis of clutch size in insects has received considerable attention in an evolutionary and behavioral context, it has received little in a demographic context. The approach taken here is to consider average clutch size as an age property of females along with the usual survival and fecundity parameters. Thus by denoting C_x as the number of clutches a female deposits at age x we then have several additional demographic relations. The sum of C_x's over all ages will yield the gross clutch rate:

$$\text{gross clutch rate} = \sum_{x=0}^{\omega} C_x \qquad (3\text{-}10a)$$

and the sum of the product $L_x C_x$ over all ages gives the net clutch rate:

$$\text{net clutch rate} = \sum_{x=0}^{\omega} L_x C_x \qquad (3\text{-}10b)$$

Table 3-12. Cumulative Parity of the Mediterranean Fruit Fly (Data from Table 3-4)

Age Class	Number of Females							Percentage of Females						
		Cumulative Egg Class							Cumulative Egg Class					
	0	1–200	201–400	401–600	601–800	>800	Total	0	1–200	201–400	401–600	601–800	>800	Total
0	48	0	0	0	0	0	48	100.0	0.0	0.0	0.0	0.0	0.0	100.0
1	48	0	0	0	0	0	48	100.0	0.0	0.0	0.0	0.0	0.0	100.0
2	48	0	0	0	0	0	48	100.0	0.0	0.0	0.0	0.0	0.0	100.0
3	39	9	0	0	0	0	48	81.3	18.8	0.0	0.0	0.0	0.0	100.0
4	16	32	0	0	0	0	48	33.3	66.7	0.0	0.0	0.0	0.0	100.0
5	6	42	0	0	0	0	48	12.5	87.5	0.0	0.0	0.0	0.0	100.0
6	1	45	0	0	0	0	46	2.2	97.8	0.0	0.0	0.0	0.0	100.0
7	1	44	1	0	0	0	46	2.2	85.7	2.1	0.0	0.0	0.0	100.0
8	1	38	7	0	0	0	46	2.2	82.6	15.2	0.0	0.0	0.0	100.0
9	1	33	12	0	0	0	46	2.2	71.7	26.1	0.0	0.0	0.0	100.0
10	1	25	20	0	0	0	46	2.2	54.3	43.5	0.0	0.0	0.0	100.0
11	1	16	29	0	0	0	46	2.2	34.8	63.0	0.0	0.0	0.0	100.0
12	1	10	33	1	0	0	45	2.2	22.2	73.3	2.3	0.0	0.0	100.0
13	1	10	33	1	0	0	45	2.2	22.2	73.3	2.3	0.0	0.0	100.0
14	0	9	32	4	0	0	45	0.0	20.0	71.1	8.9	0.0	0.0	100.0
15	0	7	29	8	0	0	44	0.0	15.9	65.9	18.2	0.0	0.0	100.0
16	0	5	29	10	0	0	44	0.0	11.4	65.9	22.7	0.0	0.0	100.0
17	0	5	25	13	1	0	44	0.0	11.4	56.8	29.5	2.3	0.0	100.0
18	0	5	22	16	1	0	44	0.0	11.4	50.0	36.4	2.3	0.0	100.0
19	0	3	19	20	2	0	44	0.0	6.8	43.2	45.5	4.5	0.0	100.0
20	0	3	14	24	3	0	44	0.0	6.8	31.8	54.5	6.8	0.0	100.0

Table 3-12. (Contd.)

21	0	2	12	23	5	0	42	0.0	4.8	28.6	54.8	11.9	0.0	100.0
22	0	1	10	23	8	0	42	0.0	2.4	23.8	54.8	19.0	0.0	100.0
23	0	1	6	27	8	0	42	0.0	2.4	14.3	64.3	19.0	0.0	100.0
24	0	1	5	24	11	1	42	0.0	2.4	11.9	57.1	26.2	2.4	100.0
25	0	1	4	24	12	1	42	0.0	2.4	9.5	57.1	28.6	2.4	100.0
26	0	1	3	21	14	1	40	0.0	2.5	7.5	52.5	35.0	2.5	100.0
27	0	1	3	20	11	1	36	0.0	2.8	8.3	55.6	30.6	2.8	100.0
28	0	1	3	14	13	4	35	0.0	2.9	8.6	40.0	37.1	11.4	100.0
29	0	1	3	9	15	6	34	0.0	2.9	8.8	26.5	44.1	17.6	100.0
	213	351	354	282	104	14	1318							

Table 3-13. Summary of Parameters and Formulae for Any Age-Specific Event at Age x, g_x

	Description	Formula
Event intensity		
	Gross rate	Σg_x
	Net rate	$\Sigma l_x g_x$
	Daily rate	$\Sigma l_x g_x / \Sigma l_x$
Event timing		
	Mean age gross schedule	$\Sigma x g_x / \Sigma g_x$
	Mean age net schedule	$\Sigma x l_x g_x / \Sigma l_x g_x$

The dividend M_x/C_x represents the number of eggs per clutch by the average female age x and

$$\text{gross eggs/clutch} = \frac{\sum_{x=0}^{\omega} M_x}{\sum_{x=0}^{\omega} C_x} \qquad (3\text{-}10c)$$

$$\text{net eggs/clutch} = \frac{\sum_{x=0}^{\omega} M_x}{\sum_{x=0}^{\omega} L_x C_x} \qquad (3\text{-}10d)$$

Gross eggs/clutch represents the average clutch size of an hypothetical cohort of females that live to the last possible age, while net eggs/clutch denotes the average clutch size of the average female in her lifetime. Additional per capita parameters for clutch relations which are easily computed using the same methods as given for M_x relations include average daily rates over the lifetime of individuals and mean ages of the gross and net clutch schedules.

Data on clutch size by age (4-day intervals) for a cohort of Mexican fruit flies is given in Table 3-14. The basic parameters are computed as follows:

$$\text{expectation of life at age } 0 = \left[4 \sum_{x=0}^{29} L_x \right]$$

$$= 75.80 \text{ days}$$

The sum of L_x's must be multiplied by a factor of 4 to account for the 4-day interval. The sums of the M_x and $L_x M_x$ columns are 1053.46 and 733.62, which give the gross and net fecundity rates, respectively. Rates that apply to the clutch data are given as

$$\text{gross clutches} = \sum_{x=0}^{29} C_x$$

$$= 162.12 \text{ clutches}$$

REPRODUCTION

$$\text{net clutches} = \sum_{x=0}^{29} L_x C_x$$

$$= 112.17 \text{ clutches}$$

$$\text{gross clutch size} = \text{gross fecundity/gross clutches}$$

$$= 1053.46/162.12$$

$$= 6.50 \text{ eggs/clutch}$$

$$\text{net clutch size} = \text{net fecundity/net clutches}$$

$$= 733.62/112.17$$

$$= 6.54 \text{ eggs/clutch}$$

$$\text{clutches/day} = \text{net clutches/expectation of life}$$

$$= 112.17/75.80$$

$$= 1.5 \text{ clutches/day}$$

Mean ages for both the gross and net clutch schedules are computed by multiplying the x-weighted schedule by 4 to account for the 4-day interval:

$$\text{mean age gross clutch} = 4 \sum_{x=0}^{\omega} x C_x \bigg/ \sum_{x=0}^{\omega} C_x$$

$$= (4) 2197.18/162.123$$

$$= 54.21$$

$$\text{mean age net clutch} = 4 \sum_{x=0}^{\omega} x L_x M_x \bigg/ \sum_{x=0}^{\omega} L_x M_x$$

$$= (4) 1283.00/112.17$$

$$= 45.75$$

Mean ages of gross and net fecundity are computed in a similar way; the former is 53.8 days, and the latter is 45.4 days. Thus for this case the mean ages of the clutch schedules are almost identical with those of the respective fecundity schedules. This will not be true for cases where clutch size changes with age.

Reproductive Life Table and Parity Progression

Techniques introduced in Chapter 2 for determining various life table properties of a cohort at a given age can be applied to reproduction to yield two important parameters, both of which are conditional on the current reproductive status (as distinct from age status) of the cohort. These are i) expectation of additional reproduction and ii) probability of advancing to the next parity class. This last measure is often referred to as parity progression (Feeney, 1983).

Table 3-14. Clutch Relations in the Mexican Fruit Fly Where C_x Denotes the Average Number of Clutches per Female at Age x (Data from Berrigan, 1987)

Age Class	Age Interval	Basic Schedules			Net Schedules			Age Weightings			
		L_x	M_x	C_x	L_xM_x	L_xC_x	xM_x	xC_x	xL_xM_x	xL_xC_x
0	0	1.00	0.00	0.000	0.00	0.000	0.00	0.00	0.00	0.00
1	1–4	1.00	0.00	0.000	0.00	0.000	0.00	0.00	0.00	0.00
2	4–8	0.93	0.00	0.000	0.00	0.000	0.00	0.00	0.00	0.00
3	8–12	0.93	2.32	0.464	2.16	0.433	6.96	1.39	6.49	1.30
4	12–16	0.93	17.28	2.712	16.12	2.530	69.12	10.85	64.49	10.12
5	16–20	0.92	36.34	5.814	33.61	5.376	181.70	29.07	168.03	26.88
6	20–24	0.90	80.77	13.630	72.69	12.267	484.62	81.78	436.16	73.60
7	24–28	0.87	68.49	9.640	59.38	8.358	479.43	67.48	415.67	58.51
8	28–32	0.87	75.00	11.000	65.03	9.537	600.00	88.00	520.20	76.30
9	32–36	0.83	82.12	11.640	68.41	9.696	739.08	104.76	615.65	87.27
10	36–40	0.83	75.16	10.400	62.61	8.663	751.60	104.00	626.08	86.63
11	40–44	0.83	65.20	9.360	54.31	7.797	717.20	102.96	597.43	85.77
12	44–48	0.83	56.04	8.640	46.68	7.197	672.48	103.68	560.18	86.37

13	48–52	0.83	43.28	6.960	36.05	5.798	562.64	90.48	468.68	75.37
14	52–56	0.83	53.52	9.600	44.58	7.997	749.28	134.40	624.15	111.96
15	56–60	0.78	49.50	6.420	38.77	5.028	742.50	96.30	581.56	75.43
16	60–64	0.64	27.85	3.920	17.87	2.515	445.60	62.72	285.85	40.23
17	64–68	0.57	37.62	5.610	21.32	3.179	639.54	95.37	362.46	54.05
18	68–72	0.48	28.53	4.980	13.79	2.408	513.54	89.64	248.30	43.34
19	72–26	0.47	30.34	5.490	14.17	2.564	576.46	104.31	269.21	48.71
20	76–80	0.43	29.57	5.450	12.82	2.363	591.40	109.00	256.37	47.25
21	80–84	0.40	48.17	7.250	19.27	2.900	1011.57	152.25	404.63	60.90
22	84–88	0.33	27.25	5.505	8.86	1.789	599.50	121.11	194.84	39.36
23	88–92	0.27	25.37	4.750	6.77	1.268	583.51	109.25	155.80	29.17
24	92–96	0.24	28.66	4.835	6.92	1.168	687.84	116.04	166.11	28.02
25	96–100	0.20	13.00	1.500	2.60	0.300	325.00	37.50	65.00	7.50
26	100–104	0.20	17.13	0.990	3.43	0.198	445.38	25.74	89.08	5.15
27	104–108	0.20	8.00	1.163	1.60	0.233	216.00	31.40	43.20	6.28
28	108–112	0.15	12.70	1.150	1.91	0.173	355.60	32.20	53.34	4.83
29	112–116	0.13	10.25	2.000	1.36	0.266	297.25	58.00	39.53	7.71
30	116–120	0.13	4.00	1.250	0.53	0.166	120.00	37.50	15.96	4.99
Total		18.95	1053.46	162.123	733.62	112.167	14164.80	2197.18	8334.45	1283.00

Suppose a cohort exists in which one individual dies after having laid one egg, one after having laid two eggs, and one after having laid three eggs. Thus using egg class rather than age class as the life scale to which survival is related, we may construct a life table as follows:

Table 3-15. Hypothetical Example of Insect Survival and Mortality Data Classified by Egg Class

Egg Class	Egg Interval	Number Dying in Interval	Fraction Dying in Interval	Fraction Surviving to Egg Class
0	0–1	0	.00	1.000
1	1–2	1	.33	.833
2	2–3	1	.33	.500
3	3–4	1	.33	.167
Total		3		2.500

Egg interval (second column from left) is used to classify live individuals that have laid i eggs but not yet $i+1$ eggs. The rightmost column corresponds to the l_x schedule from the life table and the "fraction dying in interval" column corresponds to the d_x schedule. Different perspectives on reproduction are gained by expanding on this basic theme. The application of life table methods to reproduction will be illustrated by using parity-specific data on the mosquito, *Culex pipiens* (Carey and Stenberg, unpublished). We define the following:

l_i = proportion of individuals that survive to exact parity class i

p_i = probability of surviving from parity i to $i+1$ given alive at i

q_i = probability of dying in the interval i to $i+1$ given alive at i

d_i = proportion of original individuals that die in the parity interval i to $i+1$

L_i = number of female-days lived in the parity interval i to $i+1$

T_i = number of female-days lived beyond parity i

e_i = expectation of future parity at parity i

The parity-specific life table functions are direct analogs of the age-specific life table, except that parity class i is substituted for age x. The analysis is illustrated in Table 3–16. Note from the l_i schedule that all females survived to parity 1 (i.e., they survived the nulliparous period) and 80% survived to parity 4. Nearly half survived to parity 6, and a single female survived to parity 8. The general survival pattern resembles a survival curve with low mortality at young ages and substantially higher mortality at older ages. This table also shows from the q_i schedule that the fraction of individuals at parity i that die before attaining parity $i+1$ is not independent of parity. That is, parity-specific mortality at parities 0 to 2 ranged from 5- and 20-fold less than at parities 6 to 8. Thus parity-specific mortality rates are not the same. The d_i schedule shows that the highest fraction of all deaths occurs in the interval between parities 6 and 7, and the second highest occurs between

Table 3-16. Life Table Methods Applied to Reproduction in the Mosquito, Culex pipiens (Data from Carey and Stenberg, Unpublished)

Parity, i	Number Dying in Interval i to i+1, N_i	Fraction Dying in Interval i to i+1, d_i	Fraction Living to Parity Class i, l_i	Proportion Advancing i to i+1, p_i	L_i	T_i	Expected Future Parity at Parity i, e_i
0	0	.000	1.000	1.000	1.000	5.460	5.460
1	1	.040	1.000	.960	.980	4.460	4.460
2	1	.040	.960	.958	.940	3.480	3.625
3	3	.120	.920	.870	.860	2.540	2.761
4	5	.200	.800	.750	.700	1.680	2.100
5	3	.120	.600	.800	.540	.980	1.633
6	8	.320	.480	.333	.320	.440	.917
7	3	.120	.160	.250	.100	.120	.750
8	1	.040	.040	.000	.020	.020	.500
9	0	.000	.000		.000	.000	.000
Total	25	1.000			5.460		

parities 4 and 5. Nearly 85% of all deaths occur after parity 3. The last column in Table 3-16 gives the expectation of future parity i. Note that the average newly enclosed (parity 0) *Cx. pipiens* female will experience around 5.5 reproductive events before her death, and the average individual that has attained parity 5 will experience another 1.6 reproductive events.

BIBLIOGRAPHY

Berrigan, D. (1987) Clutch size and host preference of the Mexican fruit fly, *Anastrepha ludens* (Diptera: Tephritidae). M.S. Thesis, University of California, Davis.

Böckh, R. (1986) Statistisches Jahrbuch der Stadt Berlin. Vol. 12. *Statistik des Jahres 1884*, Berlin.

Carey, J. R., P. Yang, and D. Foote (1988) Demographic analysis of insect reproductive levels, patterns and heterogeneity: Case study of laboratory strains of three Hawaiian tephritids. *Entomol. Exp. Appl.* 46:85–91.

Coale, A. J., and T. J. Trussell (1974) Model fertility schedules: Variations in the age structure of childbearing in human populations. *Popul. Index* 40:185–258.

Feeney, G. (1983) Population dynamics based on birth intervals and parity progression. *Popul. Stud.* 37:75–89.

Feeney, G., and J. A. Ross (1984) Analysing open birth interval distributions. *Popul. Stud.* 38:473–478.

Hoem, J. M. (1970) Probabilistic fertility models of the life table type. *Theor. Popul. Biol.* 1:12–38.

Karmel, P. H. (1947) The relations between male and female reproduction rates. *Popul. Stud.* 1:249—274.

Kuczynski, R. R. (1931) *Fertility and Reproduction: Methods of Measuring the Balance of Births and Deaths.* Macmillan, New York.

Menken, J. A. (1975) Biometric models of fertility. *Social Forces* 54:52–65.

Milne, J. (1977) A treatise on the valuation of annuities and assurances on lives and survivors (1815). In: D. Smith and N. Keyfitz, (eds.) Mathematical Demography: Selected Papers. Berlin and New York: Springer Verlag, pp. 27–34.

Myers, R. J. (1941) The validity and significance of male net reproduction rates. *J. Am. Stat. Assoc.* 36:275–282.

Nicander, H. (1800) In Finnish; citation in: D. Smith and N. Keyfitz (1984) *Mathematical Demography. Selected Papers.* Springer-Verlag, Berlin.

Perrin, E. B., and M. C. Sheps. (1964) Human reproduction: A stochastic process. *Biometrics* 20:28–45.

Pollard, A. H. (1948) The measurement of reproductivity. *J. Inst. Actuaries.* 74:288–305.

Potter, R. G., Jr. (1963) Birth intervals: Structure and change. *Popul. Studies* 17:155–166.

Potter, R. G. (1975) Changes of natural fertility and contraceptive equivalents. *Social Forces* 54:36–51.

Sheps, M. C., and J. A. Menken. (1973) *Mathematical Models of Conception and Birth.* University of Chicago Press, Chicago.

Westoff, C. F. (1986) Fertility in the United States. *Science* 234:554–559.

Wicksell, S. D. (1977) Nuptiality, fertility, and reproductivity (1931). In: D. Smith and N. Keyfitz, (eds.), Mathematical Demography: Selected Papers. Berlin and New York: Springer Verlag. pp. 315–322.

Wood, J. W., and M. Weinstein (1988) A model of age-specific fecundibility. *Popul. Stud.* 42:85–113.

4
Population I: Basic Concepts and Models

BACKGROUND

Population Rates

Population dynamics is concerned with population rate of change. This notion is different in an important respect from the rate of speed of a physical object like an automobile (Keyfitz, 1985). Suppose a mite population is 100 at the beginning of a week and 120 at the end of the week, then by analogy with the automobile its rate would be 20 mites per week. To become a population rate this has to be divided by the population at the start of the week. Therefore the population is growing at a rate of 20/100 or .20 per individual per week. Expressed in slightly different ways, this means that every individual in the population is increasing its percentage of the total by 20% weekly or by a factor of 1.2.

The analog of the physical rate r^* in terms of population at times t and $t + 1$, N_t, and N_{t+1} is

$$r^* = N_{t+1} - N_t \quad (= \text{amount of change}) \qquad (4\text{-}1a)$$

or

$$N_{t+1} = N_t + r^* \qquad (4\text{-}1b)$$

while the demographic rate is

$$r = (N_{t+1} - N_t)/N_t \quad (= \text{fraction or r-fold change}) \qquad (4\text{-}2a)$$

or

$$N_{t+1} = N_t(1 + r) \qquad (4\text{-}2b)$$

The rates r^* and r yield different results if projected into the future. A population of 100 mites growing by 20 per week will have grown by 40 at the end of 2 weeks and by 60 at the end of 3 weeks. However, a population growing at a rate of 20% per week will have grown by 44 at the end of 2 weeks and by 73 at the end of 3 weeks. This last rate is *geometric increase*; the former is *arithmetic increase*. The distinction between these is that an arithmetic series will have a common difference, while a geometric series will have a common ratio. For example, the series 1, 3, 5, 7, 9 is arithmetic, with the difference being 2 (i.e., $9 - 7 = 2$; $7 - 5 = 2$, etc.). The series 1, 3, 9, 27, 81 is geometric, with a common ratio of 3 (i.e., $81/27 = 3$; $27/9 = 3$; etc.). It is usually assumed that population change is geometric. For example, the mite population on cotton over a 3-week period from Carey (1983) averaged 434 mites/plant for the first week, 984 mites/plant for week 2, and 2639 mites/plant for week 3. Therefore the implied geometric growth rates are 2.3-fold (i.e., 984/434) and 2.7-fold (i.e., 2639/984) for the first and second periods, respectively.

The Balancing Equation

The simplest population model is the crude rate model, which is based on the balancing equation. This equation relates the total population this year (or month, week, day, etc.) to the total population last year. Suppose last year represents the initial population at time 0 (i.e., $t = 0$; Pop_0), then the balancing equation is given as

$$Pop_1 = Pop_0$$
$$+ \text{births}$$
$$- \text{deaths}$$
$$+ \text{in-migrants}$$
$$- \text{out-migrants} \qquad (4\text{-}3)$$

This model partials out the relative contribution of birth, death, and migration. For simplicity migration is typically neglected in demographic analysis. Since births and deaths represent totals, these terms can be re-expressed as

$$\text{births} = Pop_0 \times b$$
$$\text{deaths} = Pop_0 \times d$$

where b and d denote per capita birth and death rates, respectively. Substituting these terms into the equation yields (excluding migration terms)

$$Pop_1 = Pop_0 + (Pop_0 \times b) - (Pop_0 \times d)$$
$$= Pop_0(1 + b - d) \qquad (4\text{-}4)$$

Note that if $b - d = 0$ the population at time t will equal the population at time $t + 1$, if $b > d$ the population will increase and if $b < d$ the population will decrease. The population at time $t = 2$ (i.e., Pop_2) is determined as follows:

$$Pop_2 = Pop_1(1 + b - d)$$
$$= Pop_0(1 + b - d)(1 + b - d)$$
$$= Pop_0(1 + b - d)^2$$

The relationship for t number of time units is

$$Pop_t = Pop_0(1 + b - d)^t$$

By substituting N_t for Pop_t and λ for $(1 + b - d)$ yields

$$N_t = N_0 \lambda^t \qquad (4\text{-}5)$$

which represents the discrete (i.e., geometric) version of the most basic model of population growth. This equation can be rephrased in terms of exponentials

POPULATION I: BASIC CONCEPTS AND MODELS

by taking logs of both sides and defining $r = \ln(\lambda)$:

$$\ln N_t = t \ln(\lambda) + \ln N_0$$
$$= rt + \ln N_0 \qquad (4\text{-}6a)$$

Thus
$$N_t = N_0 e^{rt} \qquad (4\text{-}6b)$$

The crude rate model assumes that i) the population is not structured by age (homogeneity assumption); ii) birth and death rates remain fixed; and iii) closed population (i.e., no migration). The only major conceptual difference between the crude rate model and the stable population model (Lotka's equation) covered later is that of homogeneity. Age structure adds a more realistic and interesting dimension but does not change the manner of growth (i.e., geometric). Although this simple equation is unrealistic in the sense that no population can change at a constant rate forever, it is fundamental to demography in three respects. *First*, it defines population change as a compounding process. *Second*, it establishes a foundation for examining population growth patterns over short periods. That is, certain organisms do increase (or decline) geometrically under some conditions. *Third*, it provides the initial framework from which to build more complicated models, such as those with feedback terms (e.g., logistic model).

Doubling Time

The expression for geometric increase gives as the projection to time t:

$$N_t = N_0 \lambda^t \qquad (4\text{-}7)$$

where λ is the fraction of increase per unit time and N_0 is the initial number in the population. The time required for the population to double (i.e., doubling time, DT) is given by

$$\lambda^t = 2$$

or

$$t = \ln 2 / \ln \lambda$$

Since $r = \ln \lambda$, this equation reduces to

$$DT = t = \ln 2 / r \qquad (4\text{-}8)$$

For example, the doubling time for a spider mite population with an intrinsic rate of increase, $r = .2$ is $\ln 2/.2 = .69/.2 = 3.5$ days. More generally, the equation for determining the time, t, required for a population to increase by N-fold is $t = \ln(N)/r$. Thus a hypothetical mite population will triple in $[\ln(3)/.2] = 5.5$ days and increase by an order of magnitude (i.e., tenfold) in $[\ln(10)/.2] = 11.5$ days.

Growth with Subpopulations

Consider a population of mites on two separate plants. Mites on plant A increase at $\lambda_A = 1.1$ and mites on plant B at $\lambda_B = 1.4$. There are two ways to express the growth of the total population, N_t, which I refer to as cases I and II. Case I expresses the total as the sum of the two populations:

Case I
$$N_t = N_{0,A}\lambda_A^t + N_{0,B}\lambda_B^t \qquad (4\text{-}9)$$

where $N_{0,A}$ and $N_{0,B}$ denotes the numbers at time 0 in subpopulations A and B, respectively. The second case (case II) models the population resulting from the mean rate of growth between the two subpopulations:

Case II
$$N_t = N_0\lambda^t \qquad (4\text{-}10)$$

N_0 is the initial population and λ is the geometric mean of the two growth rates, λ_A and λ_B. That is, $\lambda = (\lambda_A \times \lambda_B)^{\frac{1}{2}}$. Suppose the populations begin with a single individual. That is, if $N_0 = N_{0,A} = N_{0,B} = 1.0$. Then at $t = 20$, the population for case I is

$$N_t = 1.1^{20} + 1.4^{20}$$
$$= 6.7 + 836.7$$
$$= 843.4$$

and for case II is

$$N_t = [(1.1 \times 1.4)^{\frac{1}{2}}]^{20}$$
$$= 75.0$$

Over a ten-fold difference exists between the results of the two separate projections. The point here is that separate projections of separate elements always gives a total greater than projecting the average of the two rates (Keyfitz, 1985).

Sequence of Growth Rates

Suppose the rate of increase of a mite population at time 0 is $\lambda_0 = 1.1$; at time 1, $\lambda_1 = 1.8$, and at time 2, $\lambda_2 = 1.3$. Therefore after 3 days the number in a population beginning with $N_0 = 5$ individuals will be

$$N_3 = N_0\lambda_0\lambda_1\lambda_2 \qquad (4\text{-}11)$$
$$= 5 \times 1.1 \times 1.8 \times 1.3$$
$$= 12.9$$

But note that the geometric mean of the three growth rates is 1.37 ($= 1.1 \times 1.8 \times 1.3]^{\frac{1}{3}}$) and that

$$N_3 = 5 \times 1.37^3$$
$$= 12.9$$

This one numerical example illustrates a general principle about population growth that varies through time: the resulting rate of a population for any temporal sequence is the geometric mean. Or as Keyfitz (1985) notes, this relationship shows that the numerical effect on the population total of a varying rate of growth is the same as if the arithmetic average rate applied at each moment over the time in question.

THE STABLE POPULATION MODEL

Biologists recognized the importance of birth and death rates for understanding the broad properties of insect populations long before demography was introduced to biology and ecology. But the fundamental properties of populations were not understood because there was no model in which to examine the consequences. The stable population model derived by Lotka (1907) changed this in that it provided the central conceptual and analytical tool for demographic analysis of populations that has withstood nearly eight decades of scrutiny. Its importance for understanding populations cannot be overstated since it provides the foundation upon which virtually all other population models are built.

Lopez (1961) pointed out that the objective of the stable model is to trace the dynamic characteristics of a population that starts off with an arbitrary age structure and is submitted from that moment on to a specified demographic regime. The assumptions of the stable model are basically the same as those of the crude rate model except age structure is considered. Like the crude rate model, the stable model assumes fixed birth and death rates, closed population, and only one sex.

The assumptions of the stable model bring to light some of the limitations as to the applicability of the theory to animal populations. That is, insects often migrate, sex ratios other than unity are commonplace, and birth and death rates are seldom fixed. However, as Lopez (1961) and Coale (1972) both note, the fact that a stable age distribution and the intrinsic rate of growth are long-run effects of fixed vital schedules is frequently interpreted as if those indicators had little if any predictive usefulness; i.e., that the intrinsic features of the age structure will have no occasion to emerge. This is a misinterpretation of the theory itself. The most important conclusions of the theory are twofold: *first*, that the age distribution of a population is completely determined by the history of fertility and mortality rates, and *second*, that particular schedules of birth and death rates set in motion forces that make the age structure and the rate of increase point toward an inherent steady state that is independent of initial conditions.

Coale (1972) sees stable theory as providing a population gauging tool. A particular vehicular speedometer reading means that if that speed were held constant, the vehicle bearing it would travel that distance in the designated time units. Vehicular speeds frequently change, so a particular reading will rarely be a valid predictor. Nevertheless, the speedometer is a useful instrument, and so is stable population theory if properly used.

More generally, the stable population model is useful in four respects. *First*, it provides a simple starting point for addressing fundamental questions about the *process* of population renewal. *Second*, it identifies the two most important population parameters—age distribution and growth rate—their interpendence, and their relationship with the cohort parameters of birth and death. *Third*, its output suggests a direction for population growth and a population age pattern. The sign of growth rate and the general skew of the age distribution are often more important than are the actual numbers. *Fourth*, it is capable of providing actual numbers that may serve as a frame of reference for the rate of growth, the number in each of the age classes, and the population size at a specified time. For example, the value of r is frequently used as a measure of population fitness in evolutionary biology (Charlesworth, 1980).

Derivation

Consider the simplest age-structured scalar population model consisting of two age classes

$$N_{0,t+1} = m_0 N_{0,t} + m_1 N_{1,t} \qquad (4\text{-}12a)$$

$$N_{1,t+1} = p_0 N_{0,t} \qquad (4\text{-}12b)$$

where $N_{x,t}$ and $N_{x,t+1}$ denote the numbers in the population age x at times t and t + 1, respectively; m_x is the age-specific birth rate, and p_0 is the probability of surviving from age 0 to age 1. The objective is to find the rate at which this population will increase each time step. That is, we want to find the finite rate of increase, λ. Since the rate at which the population increases also applies to the rate at which the number in each age class increases, we set λ equal to the ratio of the number in each age class at time t + 1 to the number in each age class at time t. That is,

$$\frac{N_{0,t+1}}{N_{0,t}} = \frac{N_{1,t+1}}{N_{1,t}} = \lambda \qquad (4\text{-}13a)$$

$$N_{0,t+1} = \lambda N_{0,t} \qquad (4\text{-}13b)$$

$$N_{1,t+1} = \lambda N_{1,t} \qquad (4\text{-}13c)$$

This states that the difference between the number in age class 1 at times t and t + 1 differs by a constant factor, λ. Substituting the right-hand sides of Equations 4.13b and 4.13c for $N_{0,t+1}$ and $N_{1,t+1}$ in the scalar population equations given by Equations 4.12a and 4.12b yields

$$\lambda N_{0,t} = m_0 N_{0,t} + m_1 N_{1,t} \qquad (4\text{-}14a)$$

$$\lambda N_{1,t} = p_0 N_{0,t} \qquad (4\text{-}14b)$$

Since $N_{0,t}$ can be defined in terms of $N_{1,t}$, λ, and p_0 in Equation 4.14b as

$$N_{0,t} = \frac{\lambda N_{1,t}}{p_0} \qquad (4\text{-}15)$$

the right-hand side can be substituted into the Equation 4.14a to obtain an expression containing only $N_{1,t}$:

$$\lambda \left(\frac{\lambda N_{1,t}}{p_0} \right) = m_0 \left(\frac{\lambda N_{1,t}}{p_0} \right) + m_1 N_{1,t} \qquad (4\text{-}16a)$$

or

$$0 = -\lambda^2 N_{1,t} + m_0 \lambda N_{1,t} + m_1 p_0 N_{1,t} \qquad (4\text{-}16b)$$

POPULATION I: BASIC CONCEPTS AND MODELS

Rearranging and dividing through by λ yields

$$0 = N_{1,t}(-1 + l_0 m_0 \lambda^{-1} + l_1 m_1 \lambda^{-2}) \tag{4-16c}$$

since $l_0 = 1$ and $p_0 = l_1$. Therefore the terms inside the parentheses represent the Lotka equation for two age classes:

$$1 = \sum_{x=0}^{1} \lambda^{-(x+1)} l_x m_x \tag{4-16d}$$

More generally,

$$1 = \sum_{x=0}^{\infty} \lambda^{-(x+1)} l_x m_x \tag{4-16e}$$

and by setting $\lambda = e^r$ where r is the intrinsic rate of increase, then

$$1 = \sum_{x=0}^{\infty} e^{-r(x+1)} l_x m_x \tag{4.16f}$$

which is the discrete version of the Lotka equation. Coale (1972), Keyfitz (1985), and Caswell (1989) provide several alternative methods for the derivation of this fundamental equation.

Intrinsic Rate of Increase

The intrinsic rate of increase, first introduced by Dublin and Lotka (1925), is the rate of natural increase in a closed population that has been subject to constant age-specific schedules of fertility and mortality for a long period and has converged to be a stable population (Keyfitz, 1964; Pressat, 1985). The intrinsic rate of increase is basically a special case of a crude growth rate. That is, the ratio of the total number of individuals in a population at two different points in time will yield its growth rate. The population is stable if this ratio does not change over a long time.

The exact value of the intrinsic rate of increase, r, can be determined from actual data on survival and reproduction using a procedure based on Newton's method—a numerical technique based on the formula $r_1 = r_0 - f(r)/f'(r)$, where r_0 is the original estimate of the r value; r_1 is the corrected estimate; $f(r)$ is the original Lotka equation $f(r) = [(\sum e^{-rx} L_x m_x) - 1]$; and $f'(r)$ is the first derivative of this function, or $f'(r) = (\sum x e^{-rx} L_x m_x)$. The steps for computing r include the following (see Table 4-1):

Step 1. Enter the basic data, including ages (x), in Col. 1, pivotal ages (x + .5) in Col. 2, survival (l_x) in Col. 3, midpoint survival (L_x) in Col. 4, and age-specific number of female offspring per female (m_x) in Col. 5.

Step 2. Determine the net maternity ($L_x m_x$) by multiplying Cols. 4 and 5 and enter in Col. 6.

Step 3. Estimate an r value as a starting point for progressively closer approximations. For the example presented in Table 4-1 set $r_0 = .25$ for an initial guess.

Table 4-1. Calculation of the Intrinsic Rate of Increase for the Atlantic Spider Mite, *Tetranychus turkestani* (from Carey and Bradley, 1982)

Age, x (1)	Pivotal Age, $x + .5$ (2)	Survival to Age x, l_x (3)	Mite-Days Lived x to $x+1$, L_x (4)	Female Offspring Age x, m_x (5)	Net Maternity, $L_x m_x$ (6)	First Iteration		Second Iteration		Third Iteration	
						$e^{-r_0(x+.5)}L_x m_x$ (7)	$-xe^{-r_0(x+.5)}L_x m_x$ (8)	$e^{-r_1(x+.5)}L_x m_x$ (9)	$-xe^{-r_1(x+.5)}L_x m_x$ (10)	$e^{-r_2(x+.5)}L_x m_x$ (11)	$-xe^{-r_2(x+.5)}L_x m_x$ (12)
0	0.5	1.00	1.000	.00	.000	.000	.000	.000	.000	.000	.000
1	1.5	1.00	1.000	.00	.000	.000	.000	.000	.000	.000	.000
2	2.5	1.00	1.000	.00	.000	.000	.000	.000	.000	.000	.000
3	3.5	1.00	1.000	.00	.000	.000	.000	.000	.000	.000	.000
4	4.5	1.00	1.000	.00	.000	.000	.000	.000	.000	.000	.000
5	5.5	1.00	1.000	.00	.000	.000	.000	.000	.000	.000	.000
6	6.5	1.00	1.000	.00	.000	.000	.000	.000	.000	.000	.000
7	7.5	1.00	1.000	.37	.370	.057	−.426	.038	−.285	.032	−.239
8	8.5	1.00	1.000	3.53	3.530	.422	−3.584	.268	−2.274	.220	−1.866
9	9.5	1.00	1.000	7.43	7.281	.677	−6.434	.407	−3.870	.327	−3.102
10	10.5	.96	.905	7.24	6.552	.475	−4.984	.271	−2.842	.212	−2.225
11	11.5	.85	.790	4.31	3.405	.192	−2.209	.104	−1.194	.079	−.914
12	12.5	.73	.705	6.57	4.632	.204	−2.544	.104	−1.303	.078	−.974
13	13.5	.68	.590	5.59	3.298	.113	−1.524	.055	−.740	.040	−.540
14	14.5	.50	.480	4.86	2.333	.062	−.901	.029	−.415	.020	−.296
15	15.5	.46	.420	4.70	1.974	.041	−.635	.018	−.277	.012	−.193
16	16.5	.38	.355	3.67	1.303	.021	−.347	.009	−.144	.006	−.098
17	17.5	.33	.310	4.26	1.321	.017	−.291	.007	−.114	.004	−.076
18	18.5	.29	.270	3.93	1.061	.010	−.192	.004	−.072	.003	−.046
19	19.5	.25	.240	2.48	.595	.005	−.089	.002	−.031	.001	−.020
20	20.5	.23	.220	3.83	.843	.005	−.103	.002	−.034	.001	−.021
21	21.5	.21	.200	5.09	1.018	.005	−.101	.001	−.032	.001	−.019
22	22.5	.19	.170	3.30	.561	.002	−.046	.001	−.014	.000	−.008
23	23.5	.15	.120	2.34	.281	.001	−.019	.000	−.005	.000	−.003
24	24.5	.09	.065	1.65	.107	.000	−.006	.000	−.002	.000	−.001
25	25.5	.04	.020	1.10	.022	.000	−.001	.000	−.000	.000	.000
Total				76.250	40.487	2.307	−24.434	1.318	−13.648	1.036	−10.643

POPULATION I: BASIC CONCEPTS AND MODELS

Step 4. Compute for all ages $e^{-r_0(x+.5)}L_x m_x$ (Col 7) and $xe^{-r_0(x+.5)}L_x m_x$ (Col. 8).

Step 5a. Determine the first analytical approximation for r, denoted r_1, using the equation

$$r_1 = r_0 - \left(\frac{(\text{sum Col. 7}) - 1.0}{\text{sum Col. 8}}\right)$$

$$= .25 - \frac{2.307 - 1}{-24.434}$$

$$= .303$$

Step 5b. Determine the second analytical approximation for r, denoted r_2, using the equation

$$r_2 = r_1 - \left(\frac{(\text{sum Col. 9}) - 1.0}{\text{sum Col. 10}}\right)$$

$$= .303 - \frac{1.318 - 1}{-13.648}$$

$$= .327$$

Step 5c. Determine the third analytical approximation for r, denoted r_3, using the equation

$$r_3 = r_2 - \left(\frac{(\text{sum Col. 11}) - 1.0}{\text{sum Col. 12}}\right)$$

$$= .327 - \frac{1.036 - 1}{-10.643}$$

$$= .3302$$

The value of $r = .3302$ is within three decimal places of the exact value since the adjustment at the third iteration was small. The interpretation of this value is that the Atlantic spider mite population subject to the vital rates given in Table 4-1 would eventually grow at a constant exponential rate of $r = .3302$ per individual per day or by a factor of $\lambda = 1.391$ per day.

Intrinsic Rate of Increase: Analytical Approximations

The value of r in the Lotka equation can be approximated analytically in several ways, two of which are covered here. The first method sets the variable equal to the mean age of net fecundity in the cohort, T:

$$1 = \sum_{x=0}^{\omega} e^{-rT} l_x m_x \qquad (4\text{-}17a)$$

where

$$T = \frac{\sum_{x=0}^{\omega} x l_x m_x}{\sum_{x=0}^{\omega} l_x m_x} \qquad (4\text{-}17b)$$

The denominator in this equation is the net reproductive rate, R_0. The exponential term in this equation is a constant and therefore can be brought outside the integral as

$$e^{rT} = R_0 \qquad (4\text{-}17c)$$

Therefore the analytical approximation for r is given as

$$r = \ln R_0 / T \qquad (4\text{-}17d)$$

Using the mite data given in Table 4-1, where $R_0 = 40.5$ and the mean age of reproduction is $T = 12.2$ days, we obtain an r estimation of

$$r = \frac{\ln(40.487)}{12.2}$$

$$= .303$$

This estimation differs from the exact value of $r = .3302$ by only .027.

A second method for approximating r in the Lotka equation involves two additional components—survivorship at age T, denoted l_T (using l_x notation), and the gross reproductive rate, GRR. The derivation is as follows. Both sides of the Lotka equation are first multiplied by e^{rT}:

$$e^{rT} = e^{rT} \sum_{x=0}^{\omega} e^{-rx} l_x m_x \qquad (4\text{-}18a)$$

$$e^{rT} = \sum_{x=0}^{\omega} e^{-r(x-T)} l_x m_x \qquad (4\text{-}18b)$$

Dividing both sides by l_T yields

$$e^{rT} = l_T \sum_{x=0}^{\omega} e^{-r(x-T)} (l_x/l_T) m_x \qquad (4\text{-}18c)$$

When age x equals T, then

$$\frac{l_x}{l_T} = 1.0$$

and

$$e^{-r(x-T)} = e^{-r(0)} = 1.0$$

Therefore the equation reduces to

$$e^{rT} = l_T \sum_{x=0}^{\omega} m_x \qquad (4.18d)$$

POPULATION I: BASIC CONCEPTS AND MODELS

Since the sum of m_x over all ages is the gross reproductive rate, GRR, then

$$e^{rT} = l_T \, GRR \quad (4\text{-}18e)$$

and the new analytical estimation of r is given as

$$r = \frac{\log_e l_T + \log_e GRR}{T} \quad (4\text{-}18f)$$

Using survival to age T as $L_{12} = .705$, gross reproductive rate as $GRR = 76.25$, and $T = 12.2$ in this equation gives an approximation for r of

$$r = \frac{\ln .705 + \ln 76.25}{12.2}$$

$$= .327$$

This r approximation differs for the exact r value by only .003, which is closer than the previous estimate.

Net Reproductive Rate

The net reproductive rate, denoted R_0 or NRR, is defined as the average number of female offspring that would be born to a birth cohort of females during their lifetime if they experienced a fixed pattern of age-specific birth and death rates (Pressat, 1985). The formula for net reproductive rate is

$$R_0 = \sum_{x=\alpha}^{\beta} L_x m_x \quad (4\text{-}19)$$

This parameter expresses the per generation growth rate of the population and is related to the discrete daily growth rate, λ, as given in the following example. If $R_0 = 40.5$ female mites per newborn female and mean generation time is 12.2 days, then the daily growth rate is the 12.2th root of 40.5, or $\lambda = 1.354$. To verify that R_0 is the factor by which the population has increased after the length of a generation let

$$N_t = N_0 \lambda^t$$

where N_t is the number in the population at time t, N_0 is the initial number at time 0, and λ is the finite rate of increase. If $t = T = 12.2$ days, $N_0 = 1$, and $\lambda = 1.354$, then

$$N_{12.2} = 1.354^{12.2}$$
$$= 40.5 \text{ females/newborn female}$$

Intrinsic Birth and Death Rates

The intrinsic birth rate, b, is the per capita birth rate of a population that would be reached in a closed, female population subject to fixed age-specific birth and death rates. It is the birth rate in a stable population. Its counterpart, the intrinsic death rate, d, is therefore the death rate of a population subject to the same conditions. The analytical formulae for b and d are

$$b = \frac{1}{\sum_{x=0}^{\omega} e^{-rx} L_x} \tag{4-20a}$$

$$d = b - r \tag{4-20b}$$

These intrinsic rates can be computed from data in three steps:

Step 1. Determine the sum of $L_x e^{-r(x+.5)}$ from the spider mite data (see Col. 4 in Table 4-2).

Step 2. The inverse of this sum gives the intrinsic birth rate.

$$b = 1/2.9776$$
$$= .3358$$

Table 4-2. Calculation of the Stable Age Distribution for the Atlantic Spider Mite *Tetranychus turkestani* (Data from Carey and Bradley, 1982)

Age, x (1)	Pivotal Age, x + .5 (2)	Mite-days Lived x to x + 1, L_x (3)	Exponential Weightings, $L_x \exp^{-r(x+.5)}$ (4)	Fraction in population Age x, c_x (5)	Number Age x Out of 100,000, $c_x \times 100{,}000$ (6)
0 (egg)	0.5	1.000	.8478	.2847	28472
1	1.5	1.000	.6094	.2047	20465
2	2.5	1.000	.4380	.1471	14710
3 (immature)	3.5	1.000	.3148	.1057	10573
4	4.5	1.000	.2263	.0760	7600
5	5.5	1.000	.1627	.0546	5463
6	6.5	1.000	.1169	.0393	3926
7 (adult)	7.5	1.000	.0840	.0282	2822
8	8.5	1.000	.0604	.0203	2029
9	9.5	.980	.0425	.0143	1429
10	10.5	.905	.0282	.0095	948
11	11.5	.790	.0177	.0060	595
12	12.5	.705	.0114	.0038	382
13	13.5	.590	.0068	.0023	230
14	14.5	.480	.0040	.0013	134
15	15.5	.420	.0025	.0008	84
16	16.5	.355	.0015	.0005	51
17	17.5	.310	.0010	.0003	32
18	18.5	.270	.0006	.0002	20
19	19.5	.240	.0004	.0001	13
20	20.5	.220	.0003	.0001	8
21	21.5	.200	.0002	.0001	6
22	22.5	.170	.0001	.0000	3
23	23.5	.120	.0001	.0000	2
24	24.5	.065	.0000	.0000	1
25	25.5	.020	.0000	.0000	0
Total			2.9776	1.0000	100,000

Step 3. Compute d using $d = b - r = .3358 - .3302 = .0056$ deaths per individual per day.

These rates state that each day the stable mite populations will experience .3358 births and .0056 deaths for every individual in the population, the difference of which is the intrinsic rate of increase, $r = .3302$. The intrinsic rates of birth and death are useful in three contexts. *First*, the difference $(b - d)$ equals the intrinsic rate of increase as given above. *Second*, the ratio b/d is used in extinction and stochastic population theory as a measure of the probability of each individual in a small population experiencing a single birth or a death. For example, the birth and death rates given above state that 60 births would occur per individual in a stable population for every one death (i.e., $.3358/.0056$). *Third*, their sum $(b + d)$ provides a measure of what Ryder (1975) defined as population "metabolism." This serves as a general index of the number of vital events per individual in a population. For example, the birth and death rates of the mite population state that on average .3414 vital events (births + deaths) occur daily for every individual in the population (i.e., $.3358 + .0056$).

The Stable Age Distribution

Consider a hypothetical stable population with two age classes increasing at twofold each day starting at time 0 with a single individual. Suppose further that no mortality occurs from age class 0 to 1. Then the sequence in Table 4-3 will occur for three time steps. Note that the difference between the number in age class 0 and 1 is always twofold (i.e., 4/2, 8/4, 16/8), as is the difference between the total number between two successive time steps (i.e., 6/3, 12/6, 24/12). Therefore the fraction in age class 0 will always be two-thirds and the fraction in age class 1 will always be one-third. This is the stable age distribution. Note also that the fraction of the total population in age class 1 is smaller than the fraction in age class 0 owing to growth rate and not, in this case, as a result of mortality. However, a combination of both growth rate and mortality determines the fraction of the total population in a particular age class.

Table 4-3. Hypothetical Example of a Stable Population with two age classes increasing by twofold each Time Step[1]

Age Class	Time Step			
	0	1	2	3
Age 0	2	4	8	16
Age 1	1	2	4	8
Total	3	6	12	24

[1] A constant fraction of the population is contained in each age class thus the population is at the stable age distribution (SAD).

The stable age distribution (SAD) is defined as the schedule of fractions each age class represents in the ultimate population denoted c_x. The formula for c_x is

$$c_x = \frac{e^{-rx} L_x}{\sum_{x=0}^{\omega} e^{-rx} L_x} \qquad (4\text{-}21a)$$

This formula can also be used to show the relationship between age structure and growth rate. Consider the relationship between the fraction in age class x, c_x, and the fraction in age class x + 1, c_{x+1}:

$$\frac{c_{x+1}}{c_x} = \frac{\lambda^{-(x+1)} L_{x+1}}{\lambda^{-x} L_x} \qquad (4\text{-}21b)$$

Since $l_{x+1}/l_x = p_x$, then

$$\frac{c_{x+1}}{c_x} = \frac{p_x}{\lambda} \qquad (4\text{-}21c)$$

Thus

$$\lambda c_{x+1} = p_x c_x \qquad (4\text{-}21d)$$

and

$$\lambda = \frac{p_x c_x}{c_{x+1}} \qquad (4\text{-}21e)$$

In words, growth rate of a stable population is the product of survival from age class x to x + 1 and the ratio of the fraction in age x to the fraction in age x + 1. The importance of Equation 4-21e is that three parameters must be known in order to find the fourth one. For example, survival from one age class to the next, p_x, cannot be determined from only the ratio of fractions (or numbers) in two adjacent age classes, c_x and c_{x+1}. This ratio must first be adjusted for growth rate. Likewise, growth rate, λ, cannot be determined without knowledge of the fraction in two age classes as well as their relative survivorship. These constraints apply to all population whether they are stable or not.

The stable age distribution can be computed by following these steps (see Table 4-2):

Step 1. Enter age x (Col. 1) and pivotal age (Col. 2).

Step 2. From the appropriate life table, record midpoint survival, L_x (Col. 3).

Step 3. Compute the values $L_x \exp[-r(x + .5)]$ for all x and enter into Col. 4.

Step 4. Compute the fraction of the total population in each age class, c_x by dividing each entry in Col. 4 by the sum of Col. 4 and entering in Col. 5.

Step 5. Multiply each entry in Col. 5 (i.e., the c_x's) by 100,000 to give the number in each age class of the stable population out of 100,000 individuals (Col. 6).

POPULATION I: BASIC CONCEPTS AND MODELS

Step 6. Determine the stable *stage* distribution by summing the entries in Col. 5 and Col. 6 to obtain i) the fraction of the total population in each of the three stages; and ii) the total out of 100,000 individuals in each of the stages.

The stage composition for this example of 100,000 individuals is 63,647 in the egg stage, 27,562 in the immature stage, and 8,791 in the adult stage. Thus a spider mite population subject to the birth and death rates in Table 4-2 would eventually converge to a rate of increase of .3302 per day and a stable stage distribution of 63.6% eggs, 27.6% immatures, and 8.8% adults. This stage structure is sometimes found in field populations (see Carey, 1983).

Mean Generation Time

The mean generation time, T, is defined in two ways. The first definition is the mean age of reproduction, which characterizes T as the mean interval separating the births of one generation from those of the next (Pressat, 1985). The formula for T was given in 4-17b with T = 12.2 days for the data set given in Table 4-1.

A second definition of generation time is the time required for a population to increase by a factor equal to the net reproductive rate. In other words, the time required for a newborn female to replace herself by R_0-fold. The formula here is

$$T = \frac{\ln(R_0)}{r} \tag{4-22}$$

As an example using r = .3302 and the net reproductive rate, $R_0 = 40.5$ gives a value of T = ln 40.5/.3302 = 11.2 days, which states that a population will increase by a factor of 40.5 every 11.2 days. Note that this value of generation time is one day less than generation time defined as the mean age of net reproduction from Equation 4.17b.

Reproductive Value

Reproductive value was first introduced by Fisher (1930) and thus has been used by evolutionary ecologists and geneticists primarily in theoretical contexts. It is defined as the contribution an individual age x will make to the future generation number *relative* to the contribution in population number *one newborn individual* will make over the remaining life of the female. Reproductive value is basically a sum of ratios. The sum is a *relative index* of the importance of a female's contribution to future generations but does not possess biological units. The analytical expression for reproductive value at age x, denoted v_x, is given by the equation

$$v_x = \frac{e^{r(x+1)}}{l_x \sum_{y=x}^{\omega} e^{-r(y+1)} l_y m_y} \tag{4-23a}$$

where r is the intrinsic rate of increase, and l_y and m_y denote survival to age y and reproduction at age y, respectively. The computational form of reproductive value is given by the equation

$$v_x = \frac{e^{r(x+.5)}}{L_x \sum_{y=x}^{\omega} e^{-r(y+.5)} L_y m_y} \qquad (4\text{-}23b)$$

The following steps are required for determining v_x (see Table 4-4):

Step 1. Enter age x in Col. 1, L_x in Col. 2, and m_x in Col. 3.
Step 2. Compute $e^{-r(x+.5)} L_x m_x$ and enter in Col. 4.
Step 3. Sum Col. 4 from $y = x + 1$ to ω ($= 26$).
Step 4. Compute v_x using Equation 4-22b. For example,

$$v_0 = (\text{Col. 5}) \times \frac{e^{r(0+.5)}}{L_0}$$

$$= 1.0 \times \frac{\exp(r \times .5)}{1.0}$$

$$= 1.2$$

$$V_{10} = (\text{Col. 5}) \times \frac{e^{r(10+.5)}}{L_{10}}$$

$$= \frac{e^{.33 \times 10.5}/.905}{.24}$$

$$= 8.4$$

Reproductive values for all age classes are given in Col. 6 of Table 4-4 and include the following: i) v_x increases in the prereproductive ages because survival decreases with age. Note, for example, that reproductive values are highest in ages 7 through 9 in Table 4-4. ii) After the onset of reproduction, v_x may increase or decrease depending on whether fecundity increases faster than the expectation of further life decreases. iii) Reproductive value declines to zero as the individual approaches its maximum lifespan. Charlesworth (1980) and Caswell (1989) provide important perspectives on the use of reproductive value in evolutionary biology.

POPULATION PROJECTION

Leslie Matrix

Lewis (1942) and Leslie (1945, 1948) reframed the Lotka model using matrix algebra. The importance of the resulting model, known as the Leslie matrix, is that it provides a numerical tool (as distinct from an analytical tool such

Table 4-4. Reproductive Value for Atlantic Spider Mite, *Tetranychus turkestani* (Data from Carey and Bradley, 1982)

Age, x (1)	Survival, L_x (2)	Reproduction, m_x (3)	Exponential Weighting, $e^{-r(x+.5)}L_x m_x$ (4)	Running Sums of Col. 4 $\sum_{y=x+1}^{\omega} e^{-r(y+.5)}L_y m_y$ (5)	Reproductive Value, v_x (6)
0 (egg)	1.000	0.00	0.00	1.00	1.2
1	1.000	0.00	0.00	1.00	1.6
2	1.000	0.00	0.00	1.00	2.3
3 (immature)	1.000	0.00	0.00	1.00	3.2
4	1.000	0.00	0.00	1.00	4.4
5	1.000	0.00	0.00	1.00	6.2
6	1.000	0.00	0.00	1.00	8.6
7 (adult)	1.000	0.37	0.03	0.97	11.5
8	1.000	3.53	0.21	0.76	12.5
9	.980	7.43	0.32	0.44	10.3
10	.905	7.24	0.20	0.24	8.4
11	.790	4.31	0.08	0.16	9.0
12	.705	6.57	0.07	0.08	7.5
13	.590	5.59	0.04	0.05	6.8
14	.480	4.86	0.02	0.03	6.8
15	.420	4.70	0.01	0.02	6.1
16	.355	3.67	0.01	0.01	6.3
17	.310	4.26	0.01	0.01	5.8
18	.270	3.93	0.00	0.00	5.4
19	.240	2.48	0.00	0.00	6.0
20	.220	3.83	0.00	0.00	5.3
21	.200	5.09	0.00	0.00	3.0
22	.170	3.30	0.00	0.00	1.6
23	.120	2.34	0.00	0.00	0.7
24	.065	1.65	0.00	0.00	0.2
25	.020	1.10	0.00	0.00	0.0
26	.000	0.00	0.00	0.00	
		76.25	1.00		

as the Lotka equation) for determining growth rate and age structure of populations. It is also useful for illustrating and studying the transient properties of populations as they converge to the stable state.

The Leslie matrix is of the form

$$\begin{pmatrix} F_0 & F_1 & F_2 \cdots F_{\omega-1} & F_\omega \\ P_0 & 0 & 0 \cdots 0 & 0 \\ 0 & P_1 & 0 \cdots 0 & 0 \\ \cdot & \cdot & \cdots & \cdot \\ 0 & 0 & 0 \cdots P_{\omega-1} & 0 \end{pmatrix} \begin{pmatrix} N_{1,t} \\ N_{2,t} \\ N_{3,t} \\ \cdot \\ N_{\omega,t} \end{pmatrix} = \begin{pmatrix} N_{0,t+1} \\ N_{1,t+1} \\ N_{2,t+1} \\ \cdot \\ N_{\omega,t+1} \end{pmatrix}$$

where the top row contains the birth elements $F_x = (m_x + P_x m_{x+1})/2$; the subdiagonals of $P_x = L_{x+1}/L_x$ are the period survival elements, and the vectors $N_{x,t}$ and $N_{x,t+1}$ denote the numbers at age x at times t and t + 1,

Table 4-5. Elements of a 26 × 26 Leslie Matrix for Population Projections of the Atlantic Spider Mite, *Tetranychus turkestani*

	Survival		Reproduction	
Age, x (1)	Midpoint Survival, L_x (2)	Matrix Subdiagonal, P_x (3)	Fertility, m_x (4)	Matrix Top Row, F_x (5)
0 (egg)	1.000	1.000	0.00	.000
1	1.000	1.000	0.00	.000
2	1.000	1.000	0.00	.000
3 (immature)	1.000	1.000	0.00	.000
4	1.000	1.000	0.00	.000
5	1.000	1.000	0.00	.000
6	1.000	1.000	0.00	.185
7 (adult)	1.000	1.000	0.37	1.950
8	1.000	.980	3.53	5.406
9	.980	.923	7.43	7.056
10	.905	.873	7.24	5.501
11	.790	.892	4.31	5.085
12	.705	.837	6.57	5.624
13	.590	.814	5.59	4.773
14	.480	.875	4.86	4.486
15	.420	.845	4.70	3.901
16	.355	.873	3.67	3.694
17	.310	.871	4.26	3.842
18	.270	.889	3.93	3.067
19	.240	.917	2.48	2.996
20	.220	.909	3.83	4.228
21	.200	.850	5.09	3.948
22	.170	.706	3.30	2.476
23	.120	.542	2.34	1.617
24	.065	.308	1.65	.994
25	.020	.000	1.10	.550

POPULATION I: BASIC CONCEPTS AND MODELS

respectively. Caswell (1990) provides a detailed explanation of the parameter F_x.

Construction of the Leslie matrix is completed in the following steps (see Table 4-5):

Step 1. Enter age x (Col. 1) and record the appropriate L_x values (Col. 2).
Step 2. Compute the subdiagonal elements (Col. 3) using the formula L_{x+1}/L_x. For example, $P_{10} = (L_{11}/L_{10}) = .790/.905 = 0.873$
Step 3. From the appropriate fecundity schedule, enter the m_x values in Col. 4.
Step 4. Compute the F_x values for the top row of the matrix using $F_x = (m_x + P_x m_{x+1})/2$. For example, $F_{10} = (m_{10} + P_{10} m_{11})/2 = (7.24 + .873 \times 4.31)/2 = 5.501$.

Example Iteration

A population is projected through time by first entering an initial number of individuals into one or more age classes and multiplying the Leslie matrix by the age vector through a process of one-step iteration and re-substitution. For example, consider a population with three age classes starting with $N_{0,t} = N_{1,t} = N_{2,t} = 1$ and with Leslie matrix elements of $F_0 = 0$, $F_1 = 5$, $F_1 = 3$, $P_0 = 0.8$, and $P_1 = 0.5$. Three iterations are computed as follows:

$$\begin{pmatrix} F_0 & F_1 & F_2 \\ P_0 & 0.0 & 0.0 \\ 0.0 & P_1 & 0.0 \end{pmatrix} \begin{pmatrix} N_{0,t} \\ N_{1,t} \\ N_{2,t} \end{pmatrix} = \begin{pmatrix} N_{0,t+1} \\ N_{1,t+1} \\ N_{2,t+1} \end{pmatrix}$$

Iteration 1

$$\begin{pmatrix} 0.0 & 5.0 & 3.0 \\ 0.8 & 0.0 & 0.0 \\ 0.0 & 0.5 & 0.0 \end{pmatrix} \begin{pmatrix} 1.0 \\ 1.0 \\ 1.0 \end{pmatrix} = \begin{pmatrix} 8.0 \\ 0.8 \\ 0.5 \end{pmatrix}$$

Iteration 2

$$\begin{pmatrix} 0.0 & 5.0 & 3.0 \\ 0.8 & 0.0 & 0.0 \\ 0.0 & 0.5 & 0.0 \end{pmatrix} \begin{pmatrix} 8.0 \\ 0.8 \\ 0.5 \end{pmatrix} = \begin{pmatrix} 5.5 \\ 6.4 \\ 0.4 \end{pmatrix}$$

Iteration 3

$$\begin{pmatrix} 0.0 & 5.0 & 3.0 \\ 0.8 & 0.0 & 0.0 \\ 0.0 & 0.5 & 0.0 \end{pmatrix} \begin{pmatrix} 5.5 \\ 6.4 \\ 0.4 \end{pmatrix} = \begin{pmatrix} 33.2 \\ 4.4 \\ 3.3 \end{pmatrix}$$

Note that after three iterations i) the population has grown from three individuals (i.e., one in each of three age classes) to 40.9 (i.e., 33.2 + 4.4 + 3.3), or over a thirteenfold increase; ii) the age structure has changed from one-third in each of the three classes to 92% of all individuals in age classes 0 and 1; and iii) the composition and rate of increase changed each time step.

The results of a projection with more realistic vital rates and numbers of age classes is given in Table 4-6 using the spider mite vital rates from

Table 4-6. Projection of Atlantic Spider Mite Population Starting With Three Females[1]

x	0	1	2	3	4	5	6	7	8	9
0	0	14	17	16	13	11	9	7	8	35
1	0	0	14	17	16	13	11	9	7	8
2	0	0	0	14	17	16	13	11	9	7
3	0	0	0	0	14	17	16	13	11	9
4	0	0	0	0	0	14	17	16	13	11
5	0	0	0	0	0	0	14	17	16	13
6	0	0	0	0	0	0	0	14	17	16
7	1	0	0	0	0	0	0	0	14	17
8	1	1	0	0	0	0	0	0	0	14
9	1	1	1	0	0	0	0	0	0	0
10	0	.9	.9	.9	0	0	0	0	0	0
11	0	0	.8	.8	.8	0	0	0	0	0
12	0	0	0	.7	.7	.7	0	0	0	0
13	0	0	0	0	.6	.6	.6	0	0	0
14	0	0	0	0	0	.5	.5	.5	0	0
15	0	0	0	0	0	0	.4	.4	.4	0
16	0	0	0	0	0	0	0	.4	.4	.4
17	0	0	0	0	0	0	0	0	.3	.3
18	0	0	0	0	0	0	0	0	0	.3
19	0	0	0	0	0	0	0	0	0	0
20	0	0	0	0	0	0	0	0	0	0
21	0	0	0	0	0	0	0	0	0	0
22	0	0	0	0	0	0	0	0	0	0
23	0	0	0	0	0	0	0	0	0	0
24	0	0	0	0	0	0	0	0	0	0
25	0	0	0	0	0	0	0	0	0	0
Total	3	17	35	50	63	73	82	89	96	132
N_{t+1}/N_t	5.77	1.99	1.46	1.25	1.17	1.12	1.08	1.09	1.37	1.90

Table 4-5. The population was initiated with single females in each of the three youngest adult age classes (i.e., $N_7 = N_8 = N_9 = 1.0$). Thus the total population at time t = 0 was three individuals (all adults). It is instructive to examine a "snapshot" of the population at selected intervals.

One week (t = 7 days)

In one week the population increased by a factor of 1.08 over the previous day (day 6) and contains a total of 89 individuals. Of this number, 27 are eggs, 60 are immatures, and 1.3 are adults (plus rounding errors). At this time no offspring have attained adulthood and the population growth rate is at a low point. However, the population increased by a factor of nearly 30-fold in this one-week period from the original 3 females to nearly 90 individuals.

Two weeks (t = 14 days)

In only 2 weeks the population has grown by nearly 500-fold, from 3 to 1499 individuals. In the period from week 1 to 2 it grew from 89 to 1499, or

POPULATION I: BASIC CONCEPTS AND MODELS

Table 4-6. (Contd.)

10	11	12	13	14	15	16	17	18
118	230	308	351	376	378	370	425	749
35	118	230	308	351	376	378	370	425
8	35	118	230	308	351	376	378	370
7	8	35	118	230	308	351	376	378
9	7	8	35	118	230	308	351	376
11	9	7	8	35	118	230	308	351
13	11	9	7	8	35	118	230	308
16	13	11	9	7	8	35	118	230
17	16	13	11	9	7	8	35	118
14	17	16	13	11	9	7	8	35
0	13	16	14	12	10	8	6	7
0	0	11	14	13	10	9	7	5
0	0	0	10	12	11	9	8	6
0	0	0	0	8	10	9	8	6
0	0	0	0	0	7	8	8	6
0	0	0	0	0	0	6	7	7
0	0	0	0	0	0	0	5	6
.3	0	0	0	0	0	.0	.0	4
.3	.3	0	0	0	0	.0	.0	0
.2	.2	.2	0	0	0	.0	.0	0
0	.2	.2	.2	0	0	.0	.0	0
0	0	.2	.2	.2	0	.0	.0	0
0	0	0	.2	.2	.2	.0	.0	0
0	0	0	0	.1	.1	.1	.0	0
0	0	0	0	0	.1	.1	.1	0
0	0	0	0	0	0	.0	.0	0
249	478	783	1129	1499	1869	2231	2648	3389
1.92	1.64	1.44	1.33	1.25	1.19	1.19	1.28	1.47

by 17-fold. This explosive growth was the result of high birth rates, low mortality, and rapid development. In addition the population was initiated with females at the beginning of their peak reproductive ages. The 1499 mites are divided into the three stages of 1035 eggs, 391 immatures, and 73 adults. This stage distribution of 69% eggs, 26% immatures, and 5% adults is surprisingly close to the age distribution that will eventually emerge when the population converges to stability. Middle-age adults are not present in the population at this time.

Four weeks (t = 28 days)
After 4 weeks the population increased by nearly 36,000-fold (i.e., 107,281/3) from its initial numbers and by over 70-fold over the last 2 weeks (i.e., 107,281/1499). The population stage structure is 56% eggs, 38% immatures, and 6% adults or a ratio of 9.3 eggs for every adult. All age groups are present in the population at this time.

Table 4-6. (Contd.)

x	19	20	21	22	23	24	25
0	1615	3085	4889	6711	8392	9833	11,111
1	749	1615	3085	4889	6711	8392	9833
2	425	749	1615	3085	4889	6711	8392
3	370	425	749	1615	3085	4889	6711
4	378	370	425	749	1615	3085	4889
5	376	378	370	425	749	1615	3085
6	351	376	378	370	425	749	1615
7	308	351	376	378	370	425	749
8	230	308	351	376	378	370	425
9	116	225	302	344	369	370	363
10	32	107	208	279	317	340	342
11	6	28	93	182	244	277	297
12	5	5	25	83	162	217	247
13	5	4	5	21	70	136	182
14	5	4	3	4	17	57	110
15	5	5	4	3	3	15	50
16	6	5	4	3	2	3	13
17	5	5	4	3	3	2	2
18	4	5	4	4	3	2	2
19	0	3	4	4	3	3	2
20	0	0	3	4	4	3	2
21	0	0	0	3	3	3	3
22	0	0	0	0	2	3	3
23	0	0	0	0	0	2	2
24	0	0	0	0	0	0	1
25	0	0	0	0	0	0	0
Total	4993	8054	12,898	19,534	27,817	37,502	48,430
N_{t+1}/N_t	1.61	1.60	1.51	1.42	1.35	1.29	1.27

Long term (= 40 days)

The population increased by an average factor of 1.44 per day from 3 to 6,570,943 mites, or a 2.2 million-fold increase. The population underwent an approximate 100-fold increase in the following intervals: i) from day 0 to day 10 (3 to 300); ii) from day 10 to day 23 (300 mites to 30,000 mites); and iii) from day 23 to day 38 (30,000 mites to 3,000,000). It underwent 21 doublings over the 40-day period, including a doubling from approximately 3 million on day 38 to 6 million on day 40.

This numerical perspective is important in terms of the apparent population explosion. A population doubling from 3 to 6 or from 300 to 600 will probably go unnoticed, while a population doubling from 3 million to 6 million probably will be noticed and appear to "explode" in a very short time. But the population had been increasing at a steady rate for many weeks. A schematic profile of the population structure is presented in Figure 4-1.

POPULATION I: BASIC CONCEPTS AND MODELS

Table 4-6. (Contd.)

26	27	28	29	30	31	32
13,053	17,823	28,749	48,791	78,748	117,162	161,526
11,111	13,053	17,823	28,749	48,791	78,748	117,162
9833	11,111	13,053	17,823	28,749	48,791	78,748
8392	9833	11,111	13,053	17,823	28,749	48,791
6711	8392	9833	11,111	13,053	17,823	28,749
4889	6711	8392	9833	11,111	13,053	17,823
3085	4889	6711	8392	9833	11,111	13,053
1615	3085	4889	6711	8392	9833	11,111
749	1615	3085	4889	6711	8392	9833
417	734	1583	3023	4791	6577	8224
335	385	678	1461	2790	4422	6071
298	292	336	592	1276	2436	3861
265	266	261	300	528	1138	2173
207	222	223	218	251	442	952
148	168	181	181	178	204	360
97	129	147	158	159	155	179
42	82	109	124	133	134	131
11	37	71	96	109	117	117
2	10	32	62	83	95	101
2	2	9	28	55	74	84
2	1	2	8	26	51	68
2	2	1	2	7	24	46
2	2	2	1	1	6	20
2	2	1	1	1	1	4
1	1	1	1	1	0	1
0	0	0	0	0	0	0
61,270	78,846	107,281	155,607	233,599	349,536	509,187
1.29	1.36	1.45	1.50	1.50	1.46	1.41

FUNDAMENTAL PROPERTIES OF POPULATIONS

Age Structure Transience

If two populations, one with fixed birth and death rates and the other in which these rates were changing, were to begin with a narrow age distribution, say, a single female, each would undergo a sequence of demographic changes. As current births reached maturity a new surge of births would "echo" the initial female's production of offspring, and this process would continue. In populations with fixed schedules, these echoes would be periodic and distinct initially, but they would eventually become damped. In populations with unfixed schedules the surges may or may not be distinct or periodic depending on the specific features of the changing schedules and, by definition, would never become stable. The main point is this—variation in age structure and growth rate will exist in both populations. A populations growing with fixed

Table 4-6. (Contd.)

x	33	34	35	36	37
0	209,390	260,538	322,596	418,673	590,743
1	161,526	209,390	260,538	322,596	418,673
2	117,162	161,526	209,390	260,538	322,596
3	78,748	117,162	161,526	209,390	260,538
4	48,791	78,748	117,162	161,526	209,390
5	28,749	48,791	78,748	117,162	161,526
6	17,823	28,749	48,791	78,748	117,162
7	13,053	17,823	28,749	48,791	78,748
8	11,111	13,053	17,823	28,749	48,791
9	9636	10,888	12,792	17,467	28,174
10	7591	8894	10,050	11,807	16,122
11	5300	6627	7764	8774	10,307
12	3444	4727	5911	6926	7826
13	1819	2882	3957	4948	5797
14	775	1480	2346	3221	4027
15	315	678	1295	2053	2818
16	151	266	573	1094	1735
17	115	132	232	500	955
18	102	100	115	202	436
19	90	91	89	102	180
20	77	83	83	81	94
21	62	70	75	76	74
22	39	52	60	64	64
23	14	28	37	42	45
24	2	8	15	20	23
25	0	1	1	5	6
Total	715,884	972,787	1,290,720	1,703,554	2,286,850
N_{t+1}/N_t	1.36	1.33	1.32	1.34	1.39

schedules not yet converged to a stable state may produce patterns that are virtually indistinguishable over a short period from those produced by a population subject to changing schedules.

Independence of Initial Conditions

The more remote a past age distribution becomes, the less difference its form makes to the shape of the current age distribution (Coale, 1972). The same factors that cause the transient effects of an initial age distribution to disappear from the stable population will operate for any time path of fecundity and mortality rate. After a suitably long period, the effect of an initial age distribution is swamped by the cumulative effect of the time pattern of vital rates. In short, the age distribution of any closed population is entirely determined by survival and fecundity rates of recent history (Coale, 1972). Hence, given any observed age structure, it is impossible to determine either the initial population's size or its age structure.

Table 4-6. (Contd.)

38	39	40
892,536	1,373,568	2,064,825
590,743	892,536	1,373,568
418,673	590,743	892,536
322,596	418,673	590,743
260,538	322,596	418,673
209,390	260,538	322,596
161,526	209,390	260,538
117,162	161,526	209,390
78,748	117,162	161,526
47,815	77,173	114,819
26,005	44,133	71,231
14,074	22,702	38,528
9194	12,554	20,250
6550	7696	10,508
4719	5332	6264
3524	4129	4666
2381	2978	3489
1515	2079	2600
832	1319	1811
387	740	1173
165	355	678
85	150	323
63	72	127
45	44	51
24	25	24
7	8	8
3,169,298	4,528,220	6,570,943
1.43	1.45	.00

Fertility and Mortality

Fertility differences usually have a far greater impact on current age distribution than do mortality differences (Coale, 1972). This is because the role of fertility in shaping age distributions is simpler than mortality by the fact that the differences in fertility operate in a single direction. In contrast, mortality differences have only second-order effects on age distributions. That is, a change in mortality tends to change all cohorts, implying a small effect of mortality on the immediate age distribution. When mortality changes in a gradual and monotonic fashion, the age distribution tends to adapt continuously and closely to current conditions. Thus the age structure tends to differ very little from that which would have resulted from the existence of the indefinite past of current mortality conditions.

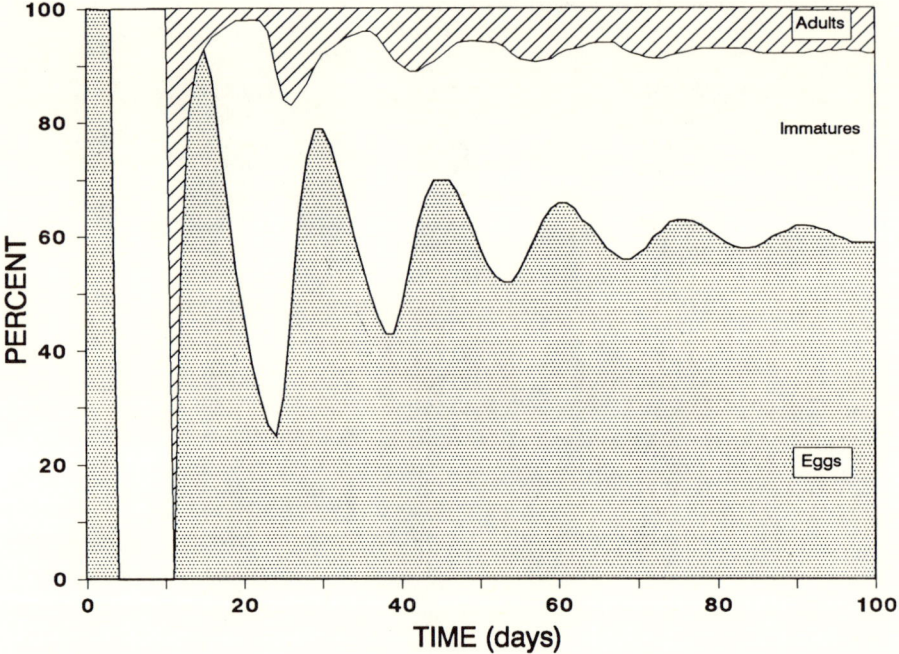

Figure 4-1. Convergence of a spider mite population stage structure to the stable age (stage) distribution.

Changing Schedules and Unchanging Age Structure

There are several instances where changing schedules are not reflected in the age distribution. One such example is in two populations characterized as exhibiting "equivalent differences" in birth and death rates. Two populations are said to have equivalent differences if a change in the birth rate in one has the same effect on age structure as a change in death rate in the other. This concept was introduced by Coale (1972) to apply to growth rate but also applies to age structure. Another case concerns only changing survivorship schedules. If survivorship in all ages were to be uniformly reduced by a constant fraction, although growth rate would decrease, there would be no change in age structure. Although extreme and hypothetical, this example helps explain why relatively inconspicuous changes in age profiles may occur despite heavy mortality differences.

Effect of Age of First Reproduction on Population Growth Rate

Lewontin (1965) was the first to ask the question, What is the effect of changes in life history parameters on the intrinsic rate of increase? There are three interrelated effects on r owing to shifts in reproductive timing that become more easily understood if viewed as effects on the components of the intrinsic rate of increase—the intrinsic birth and death rates. *First*, the shift in reproductive timing shifts the more highly fecund (young) females into different age classes. *Second*, the change in the fraction of the population in highly fecund age classes will, in turn, alter the age distribution owing to its effect on population growth rate. Therefore the effect on birth rate of a decrease in the age of first reproduction is self-reinforcing. *Third*, because a modified population growth rate causes a shift in age structure and because individuals in different age classes often have different probabilities of dying, the change in age of first reproduction changes the frequency distribution of population members in different death risk groups. In short, changes in development time after intrinsic birth and death rates in populations owing to shifts in age class weightings of vital schedules. Since peak fecundity of almost all arthropods occurs in young adults, a decrease in age of first reproduction will always increase b substantially, at least in growing populations. On the other hand, a shortening of developmental time will increase intrinsic death rate when mortality is greater in young individuals than in old individuals but decrease it when mortality is less in young than in old. This perspective also sheds light on why changes in development time have little effect on population growth rates in slowly-growing or stationary populations. The age distribution for these cases is much flatter, and therefore shifts in reproductive timing do not affect the age weightings as drastically. This helps explain the finding of Snell (1978) that Lewontin's result (i.e., r is most sensitive to changes in development time) does not hold for slowly growing populations (see Carey and Krainacker, 1988).

Speed of Convergence

Coale (1972) was one of the first demographers to examine the *process* of convergence of a population from arbitrary initial conditions to the stable form. He noted that the process of convergence is obviously an important aspect of the theory of stable populations since, if the stable form were closely approached only after several million years the stable population would be an abstract concept indeed. The determinants of the rate of convergence of a population have been examined by a number of authors, including Bernadelli (1941), Keyfitz (1972), Arthur (1981), Tuljapurkar (1982), Wachter (1984) and Kim (1985). However, Kim (1986) provided the definitive determinant of the rate of convergence. She concluded the following regarding the rate at which populations converge to the stable age distribution. *First*, for a fixed value of net reproductive rate, the speed of convergence increases as the mean of the net maternity function decreases. *Second*, for a fixed shape of the net maternity function, the speed of convergence increases as the value of net reproductive rate increases. In other words, the higher the rate of increase, given the same pattern of reproduction, the faster the population will converge to stability. *Third*, the speed of convergence of stability depends not on the shape of the net maternity function but on the shape of the stable net maternity function. The smaller the mean of the stable net maternity function, the faster the speed of convergence. Understanding the rate at which populations converge is important, not to demonstrate that populations may never attain the stable distribution by some arbitrary standard (e.g., Taylor, 1979), but rather to understand why populations in nature may depart from the stable distribution yet may still be subject to vital rates that are relatively constant.

Population Momentum

Like physical objects that have a tendency to continue moving once in motion, increasing populations have a tendency to continue growing. This is referred to as population momentum. It is the extent to which a population continues to change in size after it adopts replacement-level rates of mortality and fertility (Kim et al., 1991). As Pressat (1985) notes, the momentum of a population can be regarded as the opposite of the intrinsic rate of increase, which indicates the growth rate implicit in a set of vital schedules and independent of initial age structure. In contrast, population momentum describes the growth potential owing to age structure alone. Keyfitz (1971a,b) published the pioneering work on momentum considering an initially stable population.

BIBLIOGRAPHY

Arthur, B. (1981) Why a population converges to stability. *Am. Math. Monthly* 88:557–563.
Bernadelli, H. (1941) Population waves. *J. Burma Res. Soc.* 31(1):1–18.
Carey, J. R. (1983) Practical application of the stable age distributions: Analysis of a tetranychid mite (Acari: Tetranychidae) population outbreak. *Environ. Entomol.* 12:10–18.
Carey, J. R., and J. W. Bradley. (1982) Developmental rates, vital schedules, sex ratios and life tables of *Tetranychus urticae, T. turkestani* and *T. pacificus* (Acarina: Tetranychidae) on cotton. *Acarologia* 23:333–345.
Carey, J. R., and D. Krainacker (1988) Demographic analysis of tetranychid spider mite populations: Extensions of stable theory. *Experimental and Applied Acarology* 4:191–210.
Caswell, H. (1989) *Matrix Population Models: Construction, Analysis, and Interpretation.* Sinauer Associates, Inc., Sunderland, Massachusetts.
Charlesworth, B. (1980) *Evolution in Age-structured Populations.* Cambridge University Press, Cambridge.
Coale, A. (1972) *The Growth and Structure of Human Population.* Princeton University Press, Princeton.
Dublin, L. I., and A. J. Lotka. (1925) On the true rate of natural increase. *J. Am. Stat. Assoc.* 20:305–339.
Fisher, R. A. (1930) *The Genetical Theory of Natural Selection.* Oxford University Press, Oxford.
Keyfitz, N. (1964) The intrinsic rate, of natural increase and the dominant root of the projection matrix. *Popul. Stud.* 18:293–308.

Keyfitz, N. (1971a) Models. *Demography* 8:571–581.
Keyfitz, N. (1971b) On the momentum of population growth. *Demography* 8:71–80.
Keyfitz, N. (1972) Population Waves. In: *Population Dynamics*. T. N. E. Greville (ed.). Academic Press, New York, pp. 1–38.
Keyfitz, N, (1985) *Applied Mathematical Demography*. Springer-Verlag, New York, 441 pp.
Kim, Y. J. (1985) On the dynamics of populations with two age groups. *Demography* 22:455–468.
Kim, Y. J. (1986) Speed of convergence of stability: What matters is not net maternity function but stable net maternity function. Paper presented at the Population Association of America meetings in San Francisco.
Kim, Y. J., R. Schoen, and P. S. Sarma (1991) Momentum and the growth-free segment of a population. *Demography* 28:159–173.
Leslie, P. H. (1945) On the use of matrices in certain population mathematics. *Biometrika* 33:183–212.
Leslie, P. H. (1948) Some further notes on the use of matrices in population mathematics. *Biometrika* 35:213–245.
Lewis, E. G. (1942) On the generation and growth of a population. *Sankhya* 6:93–96.
Lewontin, R. C. (1965) Selection for colonizing ability. In: A. G. Baker and G. L. Stebbins (eds.), *The Genetics of Colonizing Species*. New York: Academic Press, pp. 77–94.
Lopez, A. (1961) *Problems in Stable Population Theory*. Office of Population Research, Princeton University, Princeton, New Jersey, 107 pp.
Lotka, A. J. (1907) Relation between birth rates and death rates. *Science* 26:21–22.
Pressat, R. (1985) *The Dictionary of Demography*. Bell and Bain, Ltd., Glasgow.
Ryder, N. B. (1975) Notes on stationary populations. *Popul. Index* 41:3–28.
Snell, T. W. (1978) Fecundity, developmental time, and population growth rate. *Oecologia* 32:119–125.
Taylor, F. (1979) Convergence to the stable age distribution in populations of insects. *Am. Natur.* 113:511–530.
Tuljapurkar, S. C. (1982) Why use population entropy? It determines the rate of covergence. *J. Math. Biol.* 13:325–337.
Wachter, K. W. (1984) Lotka's roots under rescalings. *Proc. Natl. Acad. Sci. USA* 81:3600–3604.

5

Population II: Extensions of Stable Theory

TWO-SEX MODELS

Pollack (1986) lists three different categories for questions regarding the two sexes. These are i) sexual reproduction; ii) sex ratio at birth; and iii) demography's two-sex problem, i.e., the population consequences of sex-specific birth, death, and mating. The concern here is with the last category of questions concerning the demographic implications of sex-specific differences in vital rates. The major books and papers addressing the two-sex problem in demography include those by Karmel (1948), Kendall (1949), Goodman (1953), Pollard (1973), Charlesworth (1980), Charnov (1982), Pollak (1986, 1987), and Caswell (1989).

The stable population model assumes that i) adult females produce sons at the same rate as they do daughters; and ii) female and male mortality rates are identical. These two assumptions are often violated in nature, particularly in haplodiploid arthropods such as bees, wasps, ticks, and mites. In many of these species the sex ratio is skewed as much as five to ten females for every male, and the male and female survival patterns are often different.

My objective in this section is to derive and examine models that are designed to address the following questions: i) What is the stable sex ratio when sex-specific survival schedules differ? ii) Is the stable sex ratio independent of growth rate? and iii) How will a skew towards male births shift sex ratio?

Basic Two-Sex Parameters

The starting point for analysis of a two-sex system is the tabulation of age-specific birth and death rates for the female cohort and the death rates for the male cohort. Most two-sex systems are considered *female dominant* in that births of both male and female offspring are attributed to the mothers.

The age-specific rates are denoted L_x^f and L_x^m for survival to age x of females and males, respectively, and m_x^f and m_x^m for the average respective number of female and male offspring produced by a female age x. Computation of the basic two-sex cohort parameters is illustrated using the age- and sex-specific rates for the two-spotted spider mite from Hamilton (1984) in the following steps (see Table 5-1):

Table 5-1. Age-Specific Birth and Death Rate in a Two-Sex System for *Tetranychus urticae* (Data from Hamilton, 1984; r = .3027)

Age, x (1)	Female Survival, L_x^f (2)	Female Offspring at Age x, m_x^f (3)	Net Daughter Production, $L_x^f m_x^f$ (4)	Male Survival, L_x^m (5)	Male Offspring at Age x, m_x^m (6)	Net Son Production, $L_x^f m_x^m$ (7)
0	1.00	.00	.00	1.00	.00	.00
1	1.00	.00	.00	1.00	.00	.00
2	1.00	.00	.00	1.00	.00	.00
3	1.00	.00	.00	1.00	.00	.00
4	1.00	.00	.00	1.00	.00	.00
5	1.00	.00	.00	1.00	.00	.00
6	1.00	.00	.00	.97	.00	.00
7	1.00	.00	.00	.93	.00	.00
8	1.00	.00	.00	.92	.00	.00
9	1.00	.72	.72	.92	.48	.48
10	1.00	6.03	6.03	.92	4.34	4.34
11	1.00	6.45	6.45	.91	5.95	5.95
12	.98	8.43	8.26	.90	4.20	4.12
13	.94	8.59	8.07	.89	3.72	3.50
14	.90	6.04	5.41	.88	4.51	4.04
15	.85	7.44	6.32	.87	3.81	3.24
16	.83	5.12	4.22	.86	3.93	3.24
17	.80	2.92	2.32	.85	4.47	3.55
18	.77	3.41	2.61	.83	4.00	3.06
19	.76	2.49	1.88	.81	4.89	3.69
20	.74	2.65	1.95	.79	4.48	3.29
21	.72	1.17	.84	.77	3.83	2.74
22	.68	1.44	.98	.75	2.48	1.69
23	.59	.67	.40	.74	2.06	1.22
24	.50	.50	.25	.73	2.13	1.07
25	.47	.15	.07	.71	3.63	1.71
26	.44	.54	.23	.70	2.71	1.18
27	.35	.00	.00	.70	3.00	1.04
28	.25	.00	.00	.70	1.25	.31
29	.21	.00	.00	.69	1.00	.21
30	.21	.00	.00	.68	.66	.14
31	.18	.00	.00	.67	.67	.12
32	.12	.00	.00	.66	.00	.00
33	.04	.00	.00	.64	.00	.00
34	.00	.00	.00	.62	.00	.00
35				.62		
36				.62		
37				.62		
38				.62		
39				.62		
40				.62		
41				.62		
42				.62		
43				.62		
44				.62		
45				.61		

Table 5-1. (Contd.)

Age, x (1)	Female Survival, L_x^f (2)	Female Offspring at Age x, m_x^f (3)	Net Daughter Production, $L_x^f m_x^f$ (4)	Male Survival, L_x^m (5)	Male Offspring at Age x, m_x^m (6)	Net Son Production, $L_x^f m_x^m$ (7)
46				.58		
47				.56		
48				.56		
49				.56		
50				.55		
51				.53		
52				.51		
53				.51		
54				.49		
55				.47		
56				.47		
57				.46		
58				.43		
59				.40		
60				.37		
61				.31		
62				.23		
63				.14		
64				.09		
65				.04		
66				.00		
	24.28	64.76	57.01	44.06	72.20	53.91

Step 1. Arrange female survival (L_x^f) by age x (Col. 1) in Col. 2. The sum of this column is the expectation of female life at birth, denoted e_0^f. For this case $e_0^f = 24.28$ days.

Step 2. Arrange the age-specific production of female offspring m_x^f in Col. 3. The sum of this column is the daughter gross reproductive rate, GRR_f, which equals 64.76 daughters per female. This is the number of daughters a female would lay if she lived to the maximum length of life.

Step 3. Multiply Col. 2 by Col. 3 to obtain the net maternity (Col. 4) for female offspring. The sum of this column equals 57.01 daughters per female and represents the net reproductive rate for daughters: i.e., the number of daughters the average newborn female will produce prior to her death.

Step 4. Arrange male survival by age (L_x^m) in Col. 5. The sum of this column equals the expectation of male life at birth, denoted e_0^m. For this case $e_0^m = 44.06$ days, which is the life expectancy of a newborn male.

Step 5. Arrange the production of male offspring (i.e., sons) produced by a female aged x (m_x^m) in Col. 6. The sum of this column equals the gross reproductive rate for sons, GRR_m. This rate is 72.20 sons per female. In other words, a female that lived to her maximum length of life would produce an average of around 72 sons.

Step 6. Multiply female survival (Col. 2) by female production of sons (Col. 6) to obtain the net maternity rate for sons (Col. 7). The sum of this column equals the net reproductive rate of sons or 53.91 sons per female. The interpretation of this parameter is that the average newborn female spider mite produces 53.91 sons in her lifetime.

Several aspects of these results merit comment. *First*, male survival is nearly twice that of female survival—approximately 24 days for females versus 44 days for males. *Second*, more sons would be produced by a hypothetical cohort of females that lived to the maximum length of life. That is, the gross reproductive rate for daughters was about 65 and for sons it was about 72, or around .9 daughters per son. *Third*, when reproduction is weighted by survival in the net maternity functions, the lifetime rates skew towards female offspring—around 57 daughters per average female versus about 54 sons per female or about 1.06 daughters per 1 son. This rate reflects the tendency of females to produce mostly daughters early in life and mostly sons later in life. An implication of this result is that an increase in female survival would decrease the sex ratio.

Sex Ratio at Age x

If male and female life tables are different but the rate of increase of the sexes is the same, and the ratio of male to female births is s, then the sex ratio at age x can be determined as

$$e^{-rx}L_x^f = \text{number of females age x per newborn female} \quad (5\text{-}1a)$$

$$se^{-rx}L_x^m = \text{number of males age x per newborn female} \quad (5\text{-}1b)$$

Therefore the ratio of males age x to females age x is

$$\frac{se^{-rx}L_x^m}{e^{-rx}L_x^f} = \frac{sL_x^m}{L_x^f} \quad (5\text{-}1c)$$

Thus the sex ratio at age x depends on the sex-specific life tables and the sex ratio at birth but not on the common rate of increase (Keyfitz and Beekman, 1984). The rate of increase, r, is determined from the Lotka equation using only the female rates.

Using the ratio of male and female net reproductive rates for s, or s = 0.95 (i.e., 53.91 male offspring divided by 57.01 female offspring), and survival to age 30 days for $L_{30}^f = .21$ and $L_{30}^m = .68$ yields a sex ratio of 30 days old mites as

$$SR_{30} = s\, L_{30}^m / L_{30}^f$$
$$SR_{30} = 0.95(.68)/.21$$
$$= 3.08$$

Table 5-2. Sex-Specific Survival Schedules (L_x^m and L_x^f) for the Two-Spotted Spider Mite, *T. urticae*, Weighted by Exponential Terms $e^{-r(x+.5)}$, Where $r = .3027$

Age, x	Weighted Female Survival, $e^{-r(x+.5)}L_x^f$	Weighted Male Survival, $e^{-r(x+.5)}L_x^m$
0	.860	.860
1	.635	.635
2	.469	.469
3	.347	.347
4	.256	.256
5	.189	.189
6	.140	.136
7	.103	.096
8	.076	.070
9	.056	.052
10	.042	.038
11	.031	.028
12	.022	.020
13	.016	.015
14	.011	.011
15	.008	.008
16	.006	.006
17	.004	.004
18	.003	.003
19	.002	.002
20	.001	.002
21	.001	.001
22	.001	.001
23	.000	.001
24	.000	.000
25	.000	.000
26	.000	.000
27	.000	.000
28	.000	.000
29	.000	.000
30	.000	.000
31	.000	.000
32	.000	.000
33	.000	.000
34	.000	.000
35		.000
36		.000
37		.000
38		.000
39		.000
40		.000
41		.000
42		.000
43		.000
44		.000
45		.000
46		.000
47		.000

POPULATION II: EXTENSIONS OF STABLE THEORY

Table 5-2. (Contd.)

Age, x	Weighted Female Survival, $e^{-r(x+.5)}L_x^f$	Weighted Male Survival, $e^{-r(x+.5)}L_x^m$
48		.000
49		.000
50		.000
51		.000
52		.000
53		.000
54		.000
55		.000
56		.000
57		.000
58		.000
59		.000
60		.000
61		.000
62		.000
63		.000
64		.000
65		.000
66		.000
	3.280	3.250

This states that there would exist 3.08 males 30 days old for every female mite of the same age in a stable population subject to the sex-specific rates given in Table 5-1.

Sex Ratio by Stage

The sex ratio of any two stages of the two sexes, for example, between the two adult stages, is found by summing over the adult age classes each survival schedule weighted by the term $e^{-r(x+.5)}$, dividing the sum of the male schedule by the female schedule, and multiplying this dividend by s:

$$\text{adult sex ratio} = \frac{\sum_{x=9}^{66} e^{-r(x+.5)} L_x^m}{\sum_{x=9}^{34} e^{-r(x+.5)} L_x^f} \tag{5-2}$$

This ratio can be computed using the following two steps:

Step 1. Sum the weighted schedules over the adult age classes shown in Table 5-2 from x = 9 to 34 days for females (= .205) and from x = 9 to 66 days for males (= .192).

Step 2. Using s = .95, compute the adult sex ratio as

adult sex ratio = .95(.192/.205)
= .890 adult males/adult female

This result shows that the ratio of adult males to adult females is lower than the primary sex ratio despite the fact that male longevity is nearly twice that of females. This particular ratio is the result of the interaction of two factors: i) the rapid growth rate skews the population toward the young, so that the majority of the adults of both sexes are in young age groups; ii) the females have slightly higher survival in these younger groups than do males. In short, the sex ratio for this particular example is dominated by the young.

The effect of growth rate on the adult ratio can be contrasted with the adult sex ratio of a stationary population by setting r = 0 and summing the unweighted sex-specific cohort schedules over the adult age groups and computing as before. The original sex-specific schedules were given in Table 5-1. The sum of the male schedule from x = 9 to 66 days is 35.24 and the sum of the female schedule from x = 9 to 34 days is 15.28. Therefore adult sex ratio in a mite population at zero population growth is

adult sex ratio (at r = 0) = .95(35.24/15.28)
= 2.19 adult males/adult female

Thus in slowly growing or stationary populations the effect of sex-specific longevity plays a much larger role in determining the adult sex ratio than does the growth rate.

Overall Sex Ratio

The overall sex ratio in a stable two-sex population, sometimes referred to as the *intrinsic sex ratio* (Goodman, 1953), is determined by summing the weighted sex-specific survival schedules by the exponential term as before but over all age classes. The dividend of the sum of the weighted male schedule and the sum of the female schedule multiplied by s yields the overall sex ratio of the population:

$$\text{overall population sex ratio} = \frac{\sum_{x=0}^{\omega} e^{-r(x+.5)} L_x^m}{\sum_{x=0}^{\omega} e^{-r(x+.5)} L_x^f}$$

$$= .95(3.25/3.28)$$

$$= .94 \text{ males/female (or 1.06 females/male)} \quad (5\text{-}3)$$

This result shows that the overall sex ratio is almost exactly the same as the sex ratio at birth even though differences in sex-specific survival are substantial. The overall sex ratio for the stationary case is simply the product of s and ratio of the sums of the sex-specific L_x schedules. Since these two sums are the respective sex-specific expectation of lives at birth, the overall sex ratio in a stationary population can be expressed as

$$\text{overall stationary sex ratio} = s(e_0^m/e_0^f)$$
$$= .95(44.06/24.28)$$
$$= 1.72 \text{ males per female}$$

This result states that in a stable stationary population subject to the sex-specific vital rates given in Table 5-1, there would exist 1.72 males for every female. In other words, there would exist *more* males than females in the population because of their additional longevity.

STOCHASTIC DEMOGRAPHY

General Background

Stochastic demography is defined as the theoretical and empirical study of random variation in demographic processes (Cohen, 1979). Development of stochastic demography had to wait until study of broader aspects of stochastic processes had developed sufficiently (Pollard, 1973). Early work on stochastic models include Yule (1924), McKendrick (1926), Kendall (1949), and Bartlett (1955). Understanding stochasticity in insect populations is important because insect populations are frequently subject to random environments. The inclusion of statistical uncertainty in vital rates is recognized as crucial in aiding decisions, in risk assessment in conservation and in pest management projection, and in basic inference (Tuljapurkar, 1990).

Stochasticity may be introduced into insect population models in basically two ways. The first is referred to in population biology as *demographic stochasticity*. The difference between this approach and the deterministic one is that in the deterministic model each member of the population gives birth to some tiny fraction of an individual in each small interval of time (May, 1975). But in the stochastic model only whole animals are born with specified probabilities. Goodman (1987) notes that a population goes extinct when its last member dies and that this death may be due to chance alone. Similarly, the population is reduced to its last member when its second to last member dies, and this death may also be due to chance alone. The central questions regarding demographic stochasticity have to do with the extent to which chance alone plays a role in population change. Clearly at low population levels a single event may constitute the addition or subtraction of a substantial part of the population. Therefore demographic stochasticity is especially important at low population numbers and in the likelihood of extinction (e.g., Boyce, 1977; Tuljapurkar, 1990).

The second way in which stochasticity is introduced is through *environmental stochasticity*. Environmental variation arises when the demographic rates themselves are governed by an externally driven stochastic process. Environmental randomness is due to events such as bad winters or failures of food supplies. The basic distinction between demographic and environmental stochasticity is that the former arises from fixed vital rates while the latter arises from vital rates that vary over time. Cohen (1982) provides the following example to illustrate the differences between the two and their magnitudes in the case of mortality. Suppose for a population of $N = 1,000,000$ individuals that the probability of dying within one time period is $q = .002$. The variance among populations in the number that die after one time period is $Nq(1-q) = 1996$. However, if q is a random variable with a mean of .002 and standard deviation of .0001, then the variance in the expected number of deaths among populations of a million individuals is $\text{var}(Nq) = N^2 \text{var}(q) = 10,000$. Thus the environmental variation is over five times the demographic variation.

The difference between demographic and environmental stochasticity in the case of fertility is illustrated by the following example. Suppose that the fertility of a particular age class is 1.5. In fact, only whole animals are born, so this value of fertility might arise because an individual produces either one or two offspring with equal probability. Taking this to be the case, suppose there are N individuals reproducing; the average number of offspring in 1.5N. The variance of the total number of offspring is the sum of the variances for each individual, since individuals reproduce independently: thus the variance equals $N[0.5(0.5)^2 + 0.5(0.5)^2] = 0.25N$. This is the effect of demographic stochasticity. Environmental stochasticity enters if the level of fertility for the entire age class shifts with changes in the environment. For instance, suppose that fertility is either 1.0 to 2.0 with equal probability. The average number of offspring is still 1.5N, but since all individuals experience either high or low fertility the variance is now $N[.5(.5)^2 + .5(.5)^2] = .25N^2$. This is the variance owing to environmental stochasticity. Clearly environmental stochasticity will be much more important than demographic stochasticity even for small N in that the two differ by a factor of N^2.

Environmental Variation in Vital Rates

Consider the following 2 × 2 table of age-specific fecundities:

Table 5-3. Example of Two Hypothetical Reproductive Schedules Used to Illustrate Demographic versus Environmental Stochasticity (see text)

Age	Schedule A	Schedule B
1	$m_{1A}(=30)$	$m_{1B}(=3)$
2	$m_{2A}(=20)$	$m_{2B}(=2)$

There exist several ways in which environmental stochasticity can be introduced to population growth random vital rates, in this case through random fertilities. One way is by selecting each element within an age class independently of the other age class. If either two are equally likely there will exist four possible combinations of rates with a 25% chance for each occurring on any given time step. These are

$$m_{1A} + m_{2A} = 50 \text{ eggs/day}$$
$$m_{1A} + m_{2B} = 32 \text{ eggs/day}$$
$$m_{1B} + m_{2A} = 23 \text{ eggs/day}$$
$$m_{1B} + m_{2B} = 5 \text{ eggs/day}$$
$$\sum = 110 \text{ eggs}$$

Therefore the average fecundity of a population experiencing these conditions would be $110/4 = 27.5$ with a variance of $\sigma^2 = [(50 - 27.5)^2 + \cdots + (5 - 27.5)^2]/4 = 1053/4 = 263.3$. The standard deviation is then 16.3. This case would represent *demographic stochasticity*.

A different way in which variation in vital rates can enter into the population growth rate is when rates for the two age classes are correlated. For example, suppose that both age classes have rates from the same schedule at each time, with schedules A and B being equally likely. The two possibilities are

$$\text{schedule A} = m_{1A} + m_{2A} = 50 \text{ offspring/day}$$
$$\text{schedule B} = m_{1B} + m_{2B} = 5 \text{ offspring/day}$$
$$\sum = 55 \text{ offspring}$$

Therefore the average fecundity for this case is $55/2 = 27.5$ as in the former case but the variance is $\sigma^2 = [50 - 27.5)^2 + (5 - 27.5)^2]/2 = 506.3$ and the standard deviation is 22.5.

The reason the variance is lower in the first case is that on average half the time a high rate in one age class is offset by a low rate in the other. Only 25% of the time is fecundity extremely low. In contrast, when the schedules are correlated the elements within each are either uniformly high or uniformly low and the daily variation in vital rates is much greater.

Stochastic Rate of Growth

The demographic problems surrounding these hypothetical examples are as follows: i) How will the stochastic vital rates translate to population growth rate? For example, how does a twofold difference in daily fecundity change growth rate relative to, say, a fivefold difference? ii) How will the average growth rates differ when individual *elements* are subject to stochastic processes (i.e., demographic stochasticity) versus when individual *schedules* are subject to them (i.e., environmental stochasticity)?

In a series of papers Tuljapurkar (1982a, b, 1984, 1986) and Tuljapurkar and Orzack (1980) developed the framework for examining these questions. One of the papers by Tuljapurkar (1982d) introduced an important formula for computing the stochastic rate of population increase, r_s (notation here is different than his). This parameter is not a strict analogue of the deterministic "r" inasmuch as r_s is an average of several possible rates, while r is a singular rate. The formula for this average population growth rate subject to stochastic rates is

$$r_s = r - c/2\lambda^2 \qquad (5\text{-}4)$$

where r denotes the intrinsic rate of increase for a population subjected to the Leslie matrix whose elements are the *average* of all rates for each element and c is a term that depends on how the vital rates are correlated as a result of stochasticity. When vital rates vary independently of each other as is the case for environmental stochasticity, then

$$c = \sum_{x=0}^{\infty} (\sigma^2)(d\lambda/dm_x)^2 \qquad (5\text{-}5)$$

The square root of the variance term in this equation is the standard deviation. A more complicated but similar formula applies when vital rates are correlated. It is always the case (Athreya and Karlin, 1971) that for demographic stochasticity $c = 0$.

Details for an example application of this parameter to a population of the aphid, *Aphis fabae*, are given in Table 5-4. Six age-specific fecundity levels are listed in Col. 2a through 2f, each differing by an order of magnitude from the previous one. We wish to determine average population growth rates when the population is subject to one of the following conditions: i) demographic stochasticity, where each element within an age class is subject to an equal probability of representing the fecundity level for that age class on a given day; and ii) environmental stochasticity, where the entire schedule (i.e., column) is subject to an equal probability of representing the fecundity levels for all age classes.

The computations and analysis requires the following steps:

Step 1. Determine the average and variances at each age. The average and variance for all six fecundity levels are given in Cols. 3 and 4 of Table 5-4.

Step 2. Compute the r-value based on the averages. For this we use the l_x schedule given in Col. 5 and the average m_x-value given in Col. 3 to compute the r-value for the deterministic Lotka equation whose elements are given in Col. 6. The sum of this column must equal unity. The r-value for these schedules is $r = 0.2352$ and $\lambda = 1.2652$.

Step 3. Determine the average age of fecundity in the stable population ($\hat{A}$). This quantity is simply the sum of the age-weighted elements of the Lotka equation. This sum is given at the bottom of Col. 7 as 9.84 days.

Step 4. Compute the sensitivity of λ with respect to the average m_x for all x in Col. 8. The general form of the sensitivity is $d\lambda/dm_x = e^{-r(x+1)}l_x/\hat{A}$. For example,

$$\frac{d\lambda/dm_{14}}{dm_{14}} = \frac{e^{-14(.2352)}.34}{9.84}$$

$$= .0013$$

Table 5-4. Data and Computations for Determining the Stochastic Rate of Population Growth for the Aphid, *Aphis tabae* (from Frazer, 1972) (r = .2352)

| Age, x (1) | Fecundity Levels, $(m_x's)^1$ ||||||| Mean (3) | Variance (4) | l_x (5) | $e^{-rx}l_x m_x$ (6) | $xe^{-rx}l_x m_x$ (7) | $(d\lambda/dm_x)$ (8) | $(4) \times (8)^2$ (9) |
	(2a)	(2b)	(2c)	(2d)	(2e)	(2f)							
1	.0000	.0000	.0000	.0000	.0000	.0000	.00	.00	1.00	.0000	.0000	.0803	.0000
2	.0000	.0000	.0000	.0000	.0000	.0000	.00	.00	.97	.0000	.0000	.0616	.0000
3	.0000	.0000	.0000	.0000	.0000	.0000	.00	.00	.95	.0000	.0000	.0478	.0000
4	.0000	.0000	.0000	.0000	.0000	.0000	.00	.00	.90	.0000	.0000	.0358	.0000
5	.0000	.0000	.0000	.0000	.0000	.0000	.00	.00	.88	.0000	.0000	.0276	.0000
6	.0000	.0000	.0000	.0000	.0000	.0000	.00	.00	.86	.0000	.0000	.0213	.0000
7	2.2000	.2200	.0220	.0022	.0002	.0000	.41	.65	.85	.0667	.4672	.0166	.0002
8	11.8000	1.1800	.1180	.0118	.0012	.0001	2.19	18.67	.82	.2730	2.1839	.0127	.0030
9	12.0000	1.2000	.1200	.0120	.0012	.0001	2.22	19.30	.80	.2141	1.9267	.0098	.0018
10	11.8000	1.1800	.1180	.0118	.0012	.0001	2.19	18.67	.71	.1479	1.4788	.0069	.0009
11	12.0000	1.2000	.1200	.0120	.0012	.0001	2.22	19.30	.65	.1088	1.1972	.0050	.0005
12	11.8000	1.1800	.1180	.0118	.0012	.0001	2.19	18.67	.50	.0651	.7812	.0030	.0002
13	12.2000	1.2200	.1220	.0122	.0012	.0001	2.26	19.95	.45	.0478	.6212	.0021	.0001
14	11.6000	1.1600	.1160	.0116	.0012	.0001	2.15	18.04	.34	.0271	.3799	.0013	.0000
15	10.4000	1.0400	.1040	.0104	.0010	.0001	1.93	14.50	.31	.0176	.2647	.0009	.0000
16	9.6000	.9600	.0960	.0096	.0010	.0001	1.78	12.35	.28	.0117	.1875	.0007	.0000
17	9.4000	.9400	.0940	.0094	.0009	.0001	1.74	11.85	.25	.0080	.1363	.0005	.0000
18	8.2000	.8200	.0820	.0082	.0008	.0001	1.52	9.01	.22	.0049	.0876	.0003	.0000
19	8.0000	.8000	.0800	.0080	.0008	.0001	1.48	8.58	.19	.0032	.0613	.0002	.0000
20	7.7000	.7700	.0770	.0077	.0008	.0001	1.43	7.95	.14	.0018	.0362	.0001	.0000
21	6.4000	.6400	.0640	.0064	.0006	.0001	1.19	5.49	.11	.0009	.0196	.0001	.0000
22	4.2000	.4200	.0420	.0042	.0004	.0000	.78	2.36	.07	.0003	.0068	.0000	.0000
23	4.0000	.4000	.0400	.0040	.0004	.0000	.74	2.14	.05	.0002	.0040	.0000	.0000
24	3.6000	.3600	.0360	.0036	.0004	.0000	.67	1.74	.03	.0001	.0018	.0000	.0000
25	2.2000	.2200	.0220	.0022	.0002	.0000	.41	.65	.01	.0000	.0004	.0000	.0000
26	1.9000	.1900	.0190	.0019	.0002	.0000	.35	.48	.01	.0000	.0002	.0000	.0000
27	1.0000	.1000	.0100	.0010	.0001	.0000	.19	.13	.01	.0000	.0001	.0000	.0000
28	.8400	.0840	.0084	.0008	.0001	.0000	.16	.09	.00	.0000	.0000	.0000	.0000
29	.7800	.0780	.0078	.0008	.0001	.0000	.14	.08	.00	.0000	.0000	.0000	.0000
										1.0000	9.8423		.0067

The m_x-values are given in columns 2a to 2f.

Step 5a. Compute the c-term for demographic stochasticity. We find the products over each age of the variance in Col. 4 and the square of the sensitivity in Col. 8. For example, at x = 9,

$$10 \times .0098^2 = .0018$$

The sum of Col. 9 is the c-term in Equation 5-5a, which is c = .0067.

Step 5b. Determine the average growth rate for environmental stochasticity. We use the equation

$$r_s = \frac{r - c}{2\lambda^2}$$

$$= .2352 - .0067/2(1.2652)^2$$
$$= .2352 - .0021$$
$$= .2331$$

For demographic stochasticity we simply have $r_s = r = 0.2352$.

Two general points merit comment. *First*, the value of r_s is never greater than r. The reason for this is that the age structure of a population subjected to changing vital rates is never "adjusted" to the conditions of the moment (Namboodiri, 1969). For example, a population of aphids in which the majority were in the older adult age classes would not respond very rapidly to a rapid change in vital rates for the young. *Second*, growth rate of population is affected by variance but at the same time filters and modulates variance. For example, long-term growth rate is less affected by large variance in survival in older adults for the same reason that growth rate is affected less by older adults in the deterministic case. That is, these age groups constitute a small proportion of the total population.

MULTIREGIONAL DEMOGRAPHY

Rogers (1965, 1968, 1975, 1985) is recognized as the pioneer of multiregional mathematical demography, which is concerned with the mathematical description of the changes in populations over time and space. The focus of the multiregional models is on (Feeney, 1973; Rogers, 1975): i) the *stocks* of population groups at different points in time and locations in space; ii) the *vital events* that occur among these populations; and iii) the *flows* of members of such populations across the spatial borders that delineate the constituent regions of the multiregional system. The assumptions and major conclusions of multiregional demography are analogous to those of conventional stable theory. That is, if vital rates for each subpopulation within a designated region are fixed and migration between regions is constant, the growth rate of the population will eventually become fixed and contain a stable age-by-region distribution with fixed regional shares of the overall population.

Location Aggregated by Birth Origin

The simplest population in which migration is considered consists of subpopulations without age structure in each of two regions (Rogers, 1985). Let i_A and o_A denote per capita in- and out-migration rates, respectively, for region A, and i_B and o_B denote the corresponding rates for region B. Also let b_A and d_A denote the per capita birth and death rates, respectively, for region A, and b_B and d_B denote the corresponding rates for region B. As an example, suppose that the two regions experienced in one time period the changes given in the Table 5-5.

Table 5-5. Hypothetical Data on Birth, Death and Migration for Individuals in a Two-Region Population System

Region	Initial Number	Births	Deaths	Migration In	Migration Out
A	1000	100	20	67	300
B	1000	50	10	300	67
Total	2000				

Thus the rates for each region are as shown in Table 5-6.

Table 5-6. Per Capita Birth, Death and Out-Migration Rates Computed from the Hypothetical Data Presented in Table 5-5

Region A	Region B
$b_A = 100/1000 = .10$	$b_B = 50/1000 = .05$
$d_A = 20/1000 = .02$	$d_B = 10/1000 = .01$
$o_A = 300/1000 = .30$	$o_B = 67/1000 = .067$

The rates for region A are interpreted as .10 births, .02 deaths, and .30 out-migrants per individual residing in region A per unit time. The appropriate equations that involve both the receiving and sending populations for the numbers in each of the two regions at time $t + 1$, denoted $N_{A,t+1}$ and $N_{B,t+1}$, are

$$N_{t+1,A} = N_{t,A}(1 + b_A - d_A - o_A) + o_B N_{t,B} \qquad (5\text{-}6a)$$

$$N_{t+1,B} = N_{t,B}(1 + b_B - d_B - o_B) + o_A N_{t,A} \qquad (5\text{-}6b)$$

These equations state that the populations at times $t + 1$ result from i) the increment owing to each region's natural increase; ii) the decrement owing to each region's out-migration; and iii) each region's increment owing to the in-migration from the other region. Substituting the rates given above yields

$$\begin{aligned} N_{1,A} &= 1000(1 + .10 - .02 - .30) + (.067)1000 \\ &= 1000(.78) + (.067)1000 \\ &= 780 + 66.7 \\ &= 846.7 \end{aligned}$$

$$N_{1,B} = 1000(1 + .05 - .01 - .067) + (.30)1000$$
$$= 1000(.973) + (.30)1000$$
$$= 973.3 + 300$$
$$= 1273.3$$

and

$$N_1 = N_{1,A} + N_{1,B}$$
$$= 846.7 + 1273.3$$
$$= 2120$$

The results of ten time steps are given in the Table 5-7.

Table 5-7. Results of the Two-Region Population Projection Using the Multiregional Population Parameters Given in Table 5-6

Time step	Number in Region		Percentage in Region		Total	Growth Rate (λ)
	A	B	A	B		
0	1000.0	1000.0	50.0	50.0	2000.0	1.060
1	846.7	1273.3	39.9	60.1	2120.0	1.056
2	745.3	1493.4	33.3	66.7	2238.7	1.053
3	680.9	1677.1	28.9	71.1	2358.0	1.052
4	642.9	1836.7	25.9	74.1	2479.6	1.050
5	623.9	1980.6	24.0	76.0	2604.5	1.050
6	618.7	2114.9	22.6	77.4	2733.6	1.049
7	623.6	2244.1	21.7	78.3	2867.7	1.049
8	636.0	2371.4	21.1	78.9	3007.4	1.048
9	654.2	2498.9	20.7	79.3	3153.1	1.048
10	676.8	2628.5	20.5	79.5	3305.4	
Stable percentage			19.9	80.1		

Note the following: *First*, the percentage of the total population in each of the regions stabilizes at around 20% in region A and 80% in region B. *Second*, the overall growth rate approaches a constant rate of $\lambda = 1.048$, which is also the growth rate that each of the two regions will eventually experience. The main point here is that the multiregion will eventually possess *fixed regional shares* and a *single fixed growth rate*.

Location Disaggregated by Birth Origin

The preceding model considered only the number in each of the two regions without regard to whether the individuals were born in the region of residence or not. To disaggregate the individuals within a region into those that are native born and those that are immigrants requires the following modifications. Let $_BN_{A,t}$ denote the number of individuals in region A at time t who were born in region B. That is, the left subscript denotes the region of birth. Analogous terms are $_AN_{A,t}$, $_BN_{B,t}$, and $_AN_{B,t}$. The appropriate accounting relationship for the number of individuals in region A at time t + 1 who were born in region A (i.e., $_AN_{A,t+1}$) will consist of three components:

POPULATION II: EXTENSIONS OF STABLE THEORY

i. Net number of individuals resulting from natural increase of individuals in region A who were born in region A. This component can be expressed as

$$_AN_{A,t}(1 + b_A - d_A - o_A).$$

ii. Number of births by individuals in region A who were born in region B. This is expressed as $_BN_{A,t}b_A$.

iii. Number of emigrants born in Region A who, at time t, are located in Region B. This is expressed as $o_{BA}N_{B,t}$.

Thus

$$_AN_{A,t+1} = {_AN_{A,t}}(1 + b_A - d_A - o_A) + b_{AB}N_{A,t} + o_{BA}N_{B,t} \quad (5\text{-}7a)$$

The number of individuals in region A at time t + 1 who were born in region B (i.e., $_BN_{A,t+1}$) is given by

$$_BN_{A,t+1} = {_BN_{A,t}}(1 - d_A - o_A) + o_{BB}N_{B,t} \quad (5\text{-}7b)$$

Because all births to alien migrants are added to the native population stock, this equation contains no birth rate. The analogous equations for the numbers in region B at t + 1 are

$$_BN_{B,t+1} = {_BN_{B,t}}(1 + b_B - d_B - o_B) + b_{BA}N_{B,t} + o_{AB}N_{B,t} \quad (5\text{-}8a)$$

$$_AN_{B,t+1} = {_AN_{B,t}}(1 - d_B - o_B) + o_{AA}N_{A,t} \quad (5\text{-}8b)$$

The total number in the respective regions is then

$$N_{A,t+1} = {_AN_{A,t+1}} + {_BN_{A,t+1}} \quad (5\text{-}9a)$$

$$N_{B,t+1} = {_BN_{B,t+1}} + {_AN_{B,t+1}} \quad (5\text{-}9b)$$

The set of four equations can be expressed in matrix form as

$$\begin{pmatrix} a_{11} & a_{12} & a_{12} & 0 \\ 0 & a_{22} & 0 & a_{24} \\ a_{31} & 0 & a_{33} & 0 \\ 0 & a_{42} & a_{43} & a_{44} \end{pmatrix} \begin{pmatrix} {_AN_{A,t}} \\ {_BN_{A,t}} \\ {_AN_{B,t}} \\ {_BN_{B,t}} \end{pmatrix} = \begin{pmatrix} {_AN_{A,t+1}} \\ {_BN_{A,t+1}} \\ {_AN_{B,t+1}} \\ {_BN_{B,t+1}} \end{pmatrix}$$

where

$$a_{11} = 1 + b_A - d_A - o_A$$
$$a_{12} = b_A$$
$$a_{13} = a_{24} = o_B$$
$$a_{22} = 1 - d_A - o_A$$
$$a_{31} = a_{42} = o_A$$
$$a_{32} = 1 - d_B - o_B$$
$$a_{44} = 1 + b_A - d_B - o_B$$

An example projection is presented in Table 5-8 using the same parameters in the previous model where the population was not disaggregated by birth type. Note the following: *First*, the percentage of the total population in each of the regions stabilizes at around 20% in region A and 80% in region B, as

Table 5-8. Regional Demography Disaggregated by Birth Type

Time step	Number in A			Percentage in A			Number in B			Percentage in B			Total	λ
	$_AN_A$	$_BN_A$	N_A	$_AN_A$	$_BN_A$	N_A	$_BN_B$	$_AN_B$	N_B	$_BN_B$	$_AN_B$	N_B		
0	500.0	500.0	1000.0	25.0	25.0	50.0	500.0	500.0	1000.0	25.0	25.0	50.0	2000.0	1.060
1	473.3	373.3	846.7	22.3	17.6	39.9	661.7	611.7	1273.3	31.2	28.9	60.1	2120.0	1.056
2	447.3	298.0	745.3	20.0	13.3	33.3	786.6	706.8	1493.4	35.1	31.6	66.7	2238.7	1.053
3	425.8	255.1	680.9	18.1	10.8	28.9	890.4	786.8	1677.1	37.8	33.4	71.1	2358.0	1.052
4	410.1	232.8	642.9	16.5	9.4	25.9	982.5	854.2	1836.7	39.6	34.4	74.1	2479.6	1.050
5	400.1	223.8	623.9	15.4	8.6	24.0	1068.8	911.7	1980.6	41.0	35.0	76.0	2604.5	1.050
6	395.2	223.4	618.7	14.5	8.2	22.6	1153.1	961.9	2114.9	42.2	35.2	77.4	2733.6	1.049
7	394.8	228.8	623.6	13.8	8.0	21.7	1237.4	1006.7	2244.1	43.2	35.1	78.3	2867.8	1.049
8	397.9	238.1	636.0	13.2	7.9	21.1	1323.4	1048.0	2371.4	44.0	34.8	78.9	3007.4	1.048
9	404.0	250.1	654.2	12.8	7.9	20.7	1411.9	1087.0	2498.9	44.8	34.5	79.3	3153.2	1.048
10	412.6	264.2	676.8	12.5	8.0	20.5	1503.7	1124.9	2628.5	45.5	34.0	79.5	3305.3	
19	558.5	446.8	1005.3	11.08	8.86	19.9	2531.6	1505.4	4037.0	50.2	29.9	80.1	5042.3	1.048
20	580.7	472.6	1053.3	11.0	8.9	19.9	2673.4	1557.5	4230.9	50.6	29.5	80.1	5284.2	
49	2026.0	2070.7	4096.0	9.8	10.1	19.9	11475.5	4991.1	16466.5	55.8	24.3	80.1	20563.2	1.048
50	2120.1	2173.1	4293.1	9.8	10.1	19.9	12040.2	5216.2	17256.4	55.9	24.2	80.1	21549.6	
Stable percentages				9.6	10.3	19.9				56.9	23.2	80.1		

POPULATION II: EXTENSIONS OF STABLE THEORY

in the previous case. However, within a region the percentages can be further divided into those born within the region and those born in the other region. For example, of the 19.9% of the total population in region A, 9.6% were native born and 10.3% were born in region B. *Second*, the overall growth rate approaches a constant rate of $\lambda = 1.048$, which is also the growth rate that each of the two regions will eventually experience. This is, of course, identical with the case where the birth origins were not disaggregated. The main point here is that the multiregion will eventually possess *fixed regional shares* and a *single fixed growth rate*, as in the previous case. However, each region will also possess *fixed birth origin shares*.

Age-by-Region Projection Matrix

The general configuration of a transition matrix for a two-region population with age structure is similar to a Leslie matrix in that it has a top row and a subdiagonal of non-zero elements. However, each is partitioned into groups of 2-by-2 matrices whose elements serve as either (1) the transition from *age x in region i* to *age x+1 in region j*; or (2) a birth in region i to region j. In both cases region i may be the same as region j. In other words, within one time period the individual may chose to stay put.

The projection matrix for a two-region, four-age-class population is given in Table 5-9. The notation for the survival subdiagonals is given as

$_A s_A^x$ = probability of surviving age class x within region A.
$_A s_B^x$ = probability of surviving age class x from region A to region B.
$_B s_A^x$ = probability of surviving age class x from region B to region A.
$_B s_B^x$ = probability of surviving age class x within region B.

The notation for the birth row of the matrix is

$_A b_A^x$ = number of offspring produced by individual age x in region A that stay in region A.
$_A b_B^x$ = number of offspring produced by individual age x in region A that move to region B.
$_B b_A^x$ = number of offspring produced by individual age x in region B that stay in region B.

Table 5-9. Multiregional Matrix for Two Regions, a and b, and Four Age Classes

$$\begin{pmatrix} 0 & 0 & _Ab_A^2 & _Bb_A^2 & _Ab_A^3 & _Bb_A^3 & _Ab_A^4 & _Bb_A^4 \\ 0 & 0 & _Ab_B^2 & _Bb_B^2 & _Ab_B^3 & _Bb_B^3 & _Ab_B^4 & _Bb_B^4 \\ _As_A^1 & _Bs_A^1 & 0 & 0 & 0 & 0 & 0 & 0 \\ _As_B^1 & _Bs_B^1 & 0 & 0 & 0 & 0 & 0 & 0 \\ 0 & 0 & _As_A^2 & _Bs_A^2 & 0 & 0 & 0 & 0 \\ 0 & 0 & _As_B^2 & _Bs_B^2 & 0 & 0 & 0 & 0 \\ 0 & 0 & 0 & 0 & _As_A^3 & _Bs_A^3 & 0 & 0 \\ 0 & 0 & 0 & 0 & _As_B^3 & _Bs_B^3 & 0 & 0 \end{pmatrix} \begin{pmatrix} N_{A,t}^1 \\ N_{B,t}^1 \\ N_{A,t}^2 \\ N_{B,t}^2 \\ N_{A,t}^3 \\ N_{B,t}^3 \\ N_{A,t}^4 \\ N_{B,t}^4 \end{pmatrix} = \begin{pmatrix} N_{A,t+1}^1 \\ N_{B,t+1}^1 \\ N_{A,t+1}^2 \\ N_{B,t+1}^2 \\ N_{A,t+1}^3 \\ N_{B,t+1}^3 \\ N_{A,t+1}^4 \\ N_{B,t+1}^4 \end{pmatrix}$$

Table 5-10. Matrix Elements for Four Sets of Multiregional Projections

Age	Birth Elements				Survival/Migration Elements			
	$_Ab_A$	$_Ab_B$	$_Bb_A$	$_Bb_B$	$_As_A$	$_As_B$	$_Bs_A$	$_Bs_B$
0	0	0	0	0	5/6	1/6	1/3	2/3
1	9	3	3	9	5/6	1/6	1/3	2/3
2	4	2	2	4	5/6	1/6	1/3	2/3
3	2	1	1	1	—	—	—	—

$_Bb_B^x$ = number of offspring produced by individual age x in region B that stay in region B.

The parameter values for an example multiregional projection are given in Table 5-10. Note the symmetry in birth elements between regions (i.e., $_Ab_A = {_Ab_B}$ and $_Bb_A = {_Bb_B}$) but the asymmetry in survival/migration elements. That is, region A retains 5/6 of its population but sends 1/6 to region B. However, region B retains 2/3 of its members but sends 1/3. In other words, region B is sending double the fraction of its residents to region A than region B is returning.

The results of the first five time steps of the projection starting with ten individuals in each age class for each region are presented in Table 5-11. Note the following:

1. The properties of stable theory emerge in that a stable age-by-region distribution is eventually attained with 55.3% of all residents in region A and 44.7% in region B.
2. The effect of growth within a region supersedes the effect of migration. For example, larger migration rates from region B to region A were swamped by the overall growth rate.

The addition of age structure to migration models highlights the lag between migration and its manifestation of migrants via their birth and death rates. Keyfitz (1980) commented that the matrix is like a building with a good mixing of air within each room but little circulation between rooms. We can expect that after any disturbance the within-room variation will settle down to the stable form more quickly than the between-room variation.

DEMOGRAPHIC THEORY OF SOCIAL INSECTS: THE HONEYBEE

Concept of a Superorganism

Wilson (1977) notes that the very term *colony* implies that the members are physically united, or differentiated into reproductive and sterile castes, or both. When the two conditions coexist in an advanced stage, the "society" can be viewed equally well as a superorganism or even as an organism. He points out that a dilemma exists that can be stated as follows. At what point does a society become so nearly perfect that it is no longer a society? On

Table 5-11. Iterations of the Biregional Projection Model

Time	Age in Region A					Age in Region B					Grand Total
	0	1	2	3	Sum	0	1	2	3	Sum	
0	10	10	10	10	40	10	10	10	10	40	80
1	210	12	12	12	245	210	8	8	8	235	480
2	225	244	13	13	497	195	176	8	8	383	880
3	2823	252	262	13	3349	2397	183	158	7	2745	6094
4	4211	3144	270	270	7895	3587	2291	165	150	6192	14,087
5	37,266	4693	3376	280	45,613	31,818	3419	2059	155	37,451	83,065
Stable percentage	39.5	11.6	3.3	.9	55.3	33.7	8.5	2.0	.5	44.7	100

what basis do we distinguish the extremely modified zooid of an invertebrate colony from the organ of a metazoan animal?

A superorganism consists of colonies within colonies. An individual within a colony will have gonads, somatic tissue, and a circulatory and a nervous system. Similarly, the colony of which it is a part may possess features of organization analogous to these physiological properties of the single individual (Wilson, 1977). For example, an insect colony is divided into reproductive castes (analogous to gonads) and worker castes (analogous to somatic tissue); it may exchange nutrients by trophallaxis (analogous to the circulatory system), and it may communicate a food source through certain behaviors (analogous to the nervous system). The demographic question in this context is how birth and death rates of the "organ" affect population growth rate of the "individual."

Growth Limits

Eusocial insect colonies such as honeybee colonies constitute a special kind of population in that virtually all births are directly attributable to a single individual, the queen, while all deaths are attributable to the group. They are subject to the balancing equation, like any population. But unlike population in which each female has the potential to reproduce, the contribution toward growth through births owing to the *individual* (i.e., the queen) is offset by the sum of deaths in the *group*. Consequently, colony size cannot exceed the point where the number of deaths per day in the colony is greater than the maximum daily number of offspring that a queen is capable of producing. Colony growth will only occur after all individuals that die are first replaced.

A simple model of this relationship is derived as follows. Let e_0 denote the worker expectation of life at birth. Then, $(1/e_0) = d_w$ denotes the per capita number of workers deaths in the stationary population. For example, if an individual in a colony of 1 million lives an average of 45 days (Sakagami and Fukuda, 1968; Seeley, 1985), then a total of 22,222 deaths will occur each day (i.e., 1 million divided by 45). Since there are 1440 minutes in one day, a queen must produce around 15 eggs per minute (i.e., 22,222/1440) or one egg every 4 seconds just to replace the number that die in this hypothetical colony.

The level of egg production required for queens to maintain the colony is given as

$$\text{(per capita deaths)} \times \text{(number in colony)} \qquad (5\text{-}10)$$

Let b_q denote the maximum number of eggs a queen can produce in one day. Since the expression $(1/e_0)$ gives the per capita deaths, the product of this term and the maximum number possible in a colony, denoted N^*, will give the number of eggs a queen must produce. This relationship is expressed as

$$(1/e_0)N^* = b_q \qquad (5\text{-}11a)$$

POPULATION II: EXTENSIONS OF STABLE THEORY

or

$$N^* = b_q e_0 \qquad (5\text{-}11b)$$

$$= b_q/d_w \qquad (5\text{-}11c)$$

In words, this expression states that the colony's upper size limit is equal to the product of the expectation of life of workers (days) and the maximum daily egg production rate of the queen. Mathematical relations among these three parameters are presented in Table 5-12.

Two implications emerge from this analysis. *First*, an upper limit for colony size must exist owing to demographic constraints. Even if physiological reproductive limits are not considered, a finite amount of time is needed for workers to pick up the eggs for placement in the brood chamber, as with ants and termites, or for queens to move between cells and oviposit, as is the case for bees and wasps. An example of an upper limit for birth rate in a termite queen was cited in Wilson (1971) for *Odontotermes obesus*, where the egg-lying capability was reported as 86,400 eggs/day. This is precisely one egg per second. Note from Table 5-12 that if this rate of production occurred and newborn workers lived an average of 50 days, the maximum colony size would be around 4.3 million. Put another way, a colony of 4.3 million whose individuals live an average of 50 days will experience 86,400 deaths every day. To replace these will require 86,400 births, which, in turn, will require an egg production rate for the queen of one egg per second.

Second, once the maximum colony size is attained, the only way for growth to continue is by the addition of more reproductives (i.e., queens). Multiple-queen colonies (i.e., polygyny) are common in some species of termites and ants. This is the functional equivalent of budding. Queen addition in honey bees results in (or causes) swarming. A major point here is that some *Apis mellifera* strains, particularly the Africanized strain, may undergo fission at low colony sizes, not because their queens are more fecund and their workers longer-lived, but for exactly the opposite reason. That is, their queens may be less fecund and their workers shorter-lived. The only way they can increase in number is by colony fission, not colony growth (Winston, 1979, 1987).

Table 5-12. Maximum Colony Size, N^*, Given the Expectation of Life for Workers, d_w, and Maximum Egg Production of Queens, b_q

Expectation of Worker Life (e_0) in Days	Daily Egg Production by Queen (F)[1,2]			
	100	1000	10,000	100,000
20	2000	20,000	200,000	2,000,000
50	5000	50,000	500,000	5,000,000
100	10,000	100,000	1,000,000	10,000,000
200	20,000	200,000	2,000,000	20,000,000

[1] Egg production rates of 100, 1000, 10,000, and 100,000 per day equal approximately one egg every 15 minutes, every 90 seconds, every 9 seconds, and every 0.9 seconds, respectively.

[2] This also represents the number of daily deaths that must be replaced in a colony at "equilibrium" of the specified size and member expectation of life.

Furthermore, variation in the size of colonies at swarming might reflect individual variation in reproductive rates among queens. This may partly explain why no clear relationship has been established between swarming tendencies and crowding in honey bees. That is, crowding is usually an ineluctable consequence of large size. Therefore, density dependence and numerical dependence are confounded.

Within-Colony Dynamics

The basic model assumes that a colony has only one reproducing individual (the queen in honeybees) and that she produces a fixed number, b_q, of offspring per unit time (Tuljapurkar et al., 1992). At this point all offspring are assumed to be of the same sex (workers, for honeybees), and we assume a constant death rate per unit time of d_w per nonreproductive individual. Letting N_t be the number of nonreproductive individuals in a colony at time t, the rate of change is given by

$$N_{t+1} = N_t + b_q - d_w N_t$$
$$= N_t(1 - d_w) + b_q \qquad (5\text{-}12)$$

which results in the time course of population being given by

$$N_t = N_0 e^{-d_w t} + \frac{b_q}{d_w}(1 - e^{-d_w t}) \qquad (5\text{-}13)$$

Here N_0 is some initial colony size. These equations are elementary but contain important implications.

First, note that Equation 5-12 is fundamentally different from the discrete version of the Malthusian equation because the birth rate is independent of population number. Thus population regulation is intrinsic to a colony with one reproductive individual (or indeed any fixed number of reproductives). *Second*, Equation 5-13 shows that colony size climbs towards a saturation number, which is the maximum possible colony size, distinguished by the symbol

$$N^* = \frac{b_q}{d_w} \qquad (5\text{-}14)$$

Thus N^* is re-derived and is equal to the earlier, more straightforward derivation. It represents a stable equilibrium for the population.

Several assumptions of the basic model are easily relaxed without affecting the important qualitative conclusions. *First*, suppose there are several reproductives, as in multiple-queened colonies of some social insects. One generalization of Equation 5-14 is to suppose that there is fixed number, B, of reproductives; this means that all our conclusions hold providing that we replace b_q in the equations by $b_q B$. A biologically different situation arises when a colony starts with one reproductive and then adds a second, third, etc., as the total colony population increases. The equations adapt readily to this situation since we can track numbers and age structure by allowing the birth rate, b_q, in our equations to double, triple, etc., in response to the achievement of specified threshold population numbers at which additional reproductives are added.

Second, it is expected that the birth rate of the reproductive (queen) will depend on the number N_t, usually through the availability of increased resources as more individuals are present to gather them. This situation can be described by replacing the fixed birth rate, b_q, in Equation 5-14 by a variable birth rate function, $b_q(N)$. However, the dependence of b_q on N is radically different from linear: we expect that b_q varies between some minimum level when N is very small and increases with N towards a physiologically maximal rate that cannot be exceeded however great the available resources. We can deduce the consequences of such a variable birth rate without doing any analysis. There will still be a stable limiting population size, N^*, which is approached more slowly than if b_q were fixed at its maximal value. Also, importantly, the shift in older colonies towards a younger worker population is *magnified* by the change in birth rate with N.

Swarming and Generation Time

Consider again the simple colony model in which the colony begins with some nonreproductives and one reproductive, as in Equations 5-13 and 5-14. We assume that a colony will grow until it reaches a critical *swarming size*, given by a number, N_s, of nonreproductives, and then the colony will issue a swarm. Further, when swarming occurs a number, N_0, of the nonreproductives together with one reproductive individual will stay on in the colony's physical location as an offspring colony. The remaining $(N_s - N_0)$ nonreproductives and the original reproducer will leave in a swarm and attempt to found a new colony. We define the *swarming ratio*, f, as the number of workers that leave with the swarm, N_s, relative to the maximum number possible, N^*. For example if N^* is 40,000 and the colony swarms at 30,000, then $f = .75$. This means that the swarming "trigger" is the point at which the colony has grown to three-fourths of its theoretical maximum. The equation for f is thus

$$f = N_s/N^* \tag{5-15a}$$

where N_s denotes the number that leaves with the swarm and N^* denotes the maximum number possible in the colony, given the demographic constraints of queen birth rate and worker death rate.

The *swarming fraction*, denoted $(1 - g)$, is the proportion that remain with the original colony, N_0, relative to the number in the colony at the time of swarming, N_s. For example, if the number in the colony at swarming, N_s, is 30,000, and the number that leave with the swarm is 20,000, then the number remaining is $N_0 = 10,000$. In other words, the number $([N_s - N_0] = 20,000)$ leave with the swarm.

$$1 - g = N_0/N_s \tag{5-15b}$$

We must now distinguish between a physical colony location and the collection of insects that happens to occupy that location at a particular time. The biological nature of swarming requires that the unit of demographic analysis really be the occupied physical colony location. Over time successive (offspring) groups of insects occupy the location while the insects that do not go into the offspring group always leave to try and found a new colony at a distinct physical site (Oster and Wilson, 1978; Brian, 1965, 1983). Every offspring group taking over an established physical colony site starts out with N_0 nonreproductives by our assumption above. Thus the *generation time* for an established colony may be defined as the time interval, T, in which the numbers increase from the initial N_0 to the swarming size, N_s. For our basic model we determine T by setting $t = \infty$ and $N_t = N_s$ in equation (5-13), to get

$$T = \frac{1}{d_w} \frac{\ln(1 - [1-g]f)}{1-f} \qquad (5\text{-}16)$$

This time interval, T, is a key demographic quantity because it is the basic reproductive time unit for a colony. We will shortly use this time unit to formulate the demographic dynamics of colonies.

The notable feature of Equation 5-16 is that it relates the demographic parameters that describe events within colonies to the generation time that controls the rate of creation of new colonies. This is the crux of what we mean by a *heirarchical demographic relationship*. The *first* thing that equation Equation 5-16 reveals is the dynamic consequences of biological cues that determine a decision to swarm. There is a significant difference between swarming in response to absolute density—in which case we would expect N_s to stay the same even when N* varies—and swarming a certain fraction of potential size—in which case the fraction f would stay the same even if the numbers on the right side of Equation 5-15a were to change.

The *second* thing that Equation 5-16 describes is the quantitative trade-offs between swarming size, offspring size, birth rate, and death rate in determining generation time. It is particularly interesting to consider the effect of changing death rate and birth rate on T, since we might expect these rates to be characteristically different between environments and species. All other things equal, insects with a high death rate will produce offspring colonies at a faster rate than insects with a low death rate. More generally, the individual death rate also influences colonies via its effect on the swarming ratio via Equations 5-15a and 5-15b. Thus if the swarming number, N_s, and birth rate, b_q, are the same for two species, the species with the lower death rate has a higher N* and a lower swarming fraction f; in consequence the generation time is shortened even further. It is easy to see how changes in b_q will affect T.

The *third* thing that Equation 5-16 describes is the consequence of apportionment of individuals between the offspring and parent colonies. The fraction g in Equation 5.15b is in fact an analog of offspring size for animals or seed sizes in plants, insofar as its demographic impact is concerned. If we fix the swarming fraction g, we find that T is shortened by decreasing the swarming ratio f. For fixed swarming ratio f we have decreasing T as swarming fraction g decreases. Thus a colony can reproduce very frequently by

POPULATION II: EXTENSIONS OF STABLE THEORY

swarming at low swarm numbers and by putting most of the colony's individuals in the offspring colony. The costs of such a swarming strategy lie in the survival probability of swarms that leave.

Colony Demography

We are now in a position to move up in our hierarchy and consider the formation and dynamics of colonies. According to the assumptions of our basic model, we focus on a colony as a physical site occupied by a particular group of insects. Every such site starts out as a new location chosen by a swarm that has issued from some other established site. Such a swarm, in our model, contains $(N_s - N_0)$ insects at the time of formation, and we suppose that there is a probability, p, that such a swarm will find a physical colony and then increase in number to N_s, at which time it will issue a swarm. Subsequent events in this newly established colony follow a cycle of offspring colonies' growing until they swarm. Figure 5-1 shows the "life cycle" graph of a colony. We assume that an offspring colony (whose initial size is always N_0) has a probability, p, of reaching swarming size. The time between establishment of a site and the issuance of the first swarm is T_1, computed just as in Equation 5-16 but allowing for the different initial number, so that

$$T_1 = \frac{1}{d_w} \ln \frac{(1-gf)}{1-f} \qquad (5\text{-}17)$$

Once the first swarm has been issued the time between successive swarming events is the colony generation time, T, from Equation 5-16.

We can now use Figure 5-1 and the basic parameters to analyze the demographic dynamics of colonies, viewed as self-renewing aggregates. This is the top level of our heirarchy. Thus the first circle in Figure 5-1 is a newborn colony with survival probability p* until the first "reproductive event," which is the second circle. At each circle, the colony "reproduces" by

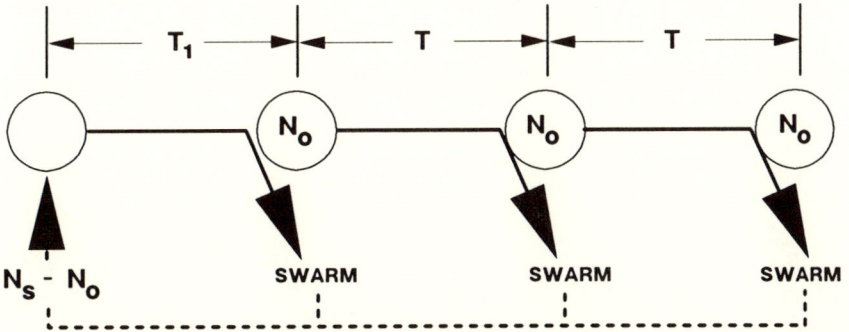

Figure 5-1. Honeybee swarming scheme as described by model. N_s denotes the number at which the colony swarms and N_0 the number that remain at the physical colony site (parent colony). Therefore the "newborn" colony depicted in the first circle starts with $(N_s - N_0)$ individuals.

swarming with a "fecundity" of 1. Standard demographic reasoning can now be applied to Figure 5-1. Specifically, we use Lotka's renewal equation with a fecundity function that has pulsed production of single offspring at each of the time T_1, $T_1 + T$, $T_1 + 2T$, etc. The survivorship function at these times has the value p*, p*p, p*p², etc. The renewal equation then yields the following equation for the long-run exponential growth rate, r, of colonies:

$$e^{rT} = p + p^* e^{r(T - T_1)} \qquad (5\text{-}18)$$

It is customary to write λ for the quantity e^{rT}; this λ is the growth rate per generation time, T. Equation 5-18 is the key result here and it completes the integration of within-colony events into the dynamics of a population viewed as an assemblage of colonies. Note that if g = .5—or, in words, if a swarming colony always puts half its numbers into offspring—then $T = T_z$ and Equation 5-18 simplifies to $\lambda = p + p^*$.

Application

As a specific example, suppose the daily birth rate of a queen, b_q, is 1000 and the expectation of life of newborn workers is 50 days, yielding a daily death rate of $d_w = .02$. The maximum colony size is then

$$N^* = 1000/.02$$
$$= 50{,}000 \text{ workers/colony}$$

Suppose that the colony swarms at $N_s = 35{,}000$; then the threshold fraction, f, is

$$f = N_s/N^*$$
$$= 35{,}000/50{,}000$$
$$= .70$$

This states that the colony swarms at 70% of its maximum possible size. Suppose also that of the 35,000 in the colony at the time of swarming 20,000 leave with the old queen. Then the swarming fraction, (1 − g), is

$$(1 - g) = N_0/N_s$$
$$= 15{,}000/35{,}000$$
$$= .43$$

In other words, 43% of the colony remains with the physical colony at the time of swarming and $g = (1 - .43) = .57$.

The generation time is thus given by

$$T = \frac{1}{d_w} \ln\left[\frac{1 - (1-g)f}{1-f} \right]$$

$$= \frac{1}{.02} \ln\left[\frac{1-(1-.57).70}{1-.70}\right]$$

$$= 42.3 \text{ days}$$

The parameter T is interpreted as interswarm interval in days of the original colony. The first generation of the founding colony is given by

$$T_1 = \frac{1}{d_w} \ln\left[\frac{(1-gf)}{1-f}\right]$$

$$= \frac{1}{.02} \ln\left[\frac{1-(.57)(.70)}{1-.7}\right]$$

$$= 34.7 \text{ days}$$

Setting the probability of a newborn colony's surviving to produce its first swarm as $p^* = .4$ and the probability of survival to subsequent swarms as $p = .8$, the rate of daily increase per colony is described by the parameterized renewal formula as

$$e^{rT} = p + p^* e^{r(T-T_1)}$$
$$e^{42.3} = .8 + .4[e^{r(42.3-34.7)}]$$
$$e^{42.3} = .8 + .4 e^{7.6}$$

The value of r that satisfies this is $r = .0046$. This is interpreted as the per colony rate of daily increase. That is, each colony produces another colony at a rate of .46% per day. The doubling time of colonies is therefore

$$DT = \frac{\ln 2}{r}$$

$$= \frac{.69}{.0046}$$

$$= 151 \text{ days}$$

The per generation growth rate is $\lambda = e^{rT} = e^{(.0046)(42.3)} = 1.2$ newborn colonies per colony per generation.

Some overall generalizations include the following: *First*, a change in the expectation of life of workers has a simple linear effect on the generation time. *Second*, the major determinant of generation time is the threshold fraction—the fraction of the theoretical maximum that the colony swarms. *Third*, the swarming fraction (i.e., the fraction of the total colony that stays with the physical colony) is of secondary importance since the colony will always begin well below the theoretical maximum where growth rate is high. *Fourth*, generation time (T) is independent of birth rate (b_q), though colony size is not. Since r is dependent on T but T is independent of the birth rate of the queen (b_q), then r is independent of the b_q. Therefore worker bee death rate (d_w) is the primary life history trait of individuals that determines colony growth rate. The reason for this is that the lower the death rate of workers, the greater the fraction of new births that will contribute to growth rate and not simply to replacement.

THE UNITY OF DEMOGRAPHIC POPULATION MODELS

A large and significant body of theory, methods, and applications in demography is concerned with the transitions that individuals experience during their lifetime as they pass from one state of existence to another (Land and Rogers, 1982). For example, from being alive to being dead, from being age 20 to being age 21, or from living in one region to living in another region. These refer to the "state of existence," while the transition rates refer to the "rules of transition." By subjecting the "states of existence" to specified "rules of transition," we obtain a set of properties that apply to a wide range of demographic population models. This common thread is erogidicity—the property of having the present state of a population independent of its makeup in the remote past (Pressat, 1985). The unifying ergodic threads for each of a number of different demographic population models are described (or redescribed) as follows:

Leslie Matrix Models

The *age structure* (SAD) and *growth rate* (r) of two populations beginning with different initial conditions will be fixed and identical when projected into the distant future if they are both subject to the same regime of constant age-specific birth and death rates. This property of the deterministic case is referred to as *strong ergodicity*.

Models Structured by Stage or Size

The *size or stage structure* and *growth rate* of two populations beginning with different initial conditions will be fixed and identical when projected into the distant future if they are both subject to the same regime of constant size- or stage-specific birth, death, and transition rates (Werner and Caswell, 1977; Caswell and Werner, 1978; Caswell, 1989).

Stationary Population Models

The *age structure* of two stationary populations that begin with different initial conditions and are projected into the distant future will be identical if they are subject to the same life table and *replacement-level fertility* (Ryder, 1973, 1975).

Stochastically Varying Vital Rates

The *age structure* and *long-term growth rate* of two populations structured by age and beginning with different initial conditions will be similar when projected into the distant future if they are both subject to the same *set* of age-specific birth and death *probabilities*. This property of stochastic models is referred to as *weak stochastic ergodicity* (Cohen, 1979).

Deterministically Varying Vital Rates

The age structure and long-term growth rate of two populations structured by age and beginning with different initial conditions will follow identical trajectories when projected into the distant future if they are both subject to the same *deterministic sequence* of age-specific birth and death rates (Tuljapurkar, 1987a, b). In this case an external cycle is superimposed on an intrinsic or internal demographic one (i.e., the generation cycle). This property is referred to as *weak deterministic ergodicity*.

Two-Sex Model

The *age structure, growth rate*, and *sex composition* (intrinsic sex ratio) of two populations structured by age and sex beginning with different initial conditions and subject to the same regime of constant sex-specific birth and death rates will be constant and identical when projected into the distant future (Goodman, 1953).

Multiregional Models

The *region-specific age structure, regional shares* of the total population, and *overall population growth rate* of two populations identically divided into regions that begin with different initial conditions will be constant and identical when projected into the future if they are subject to the same fixed regime of age-specific birth, death, and migration rates (Rogers, 1975).

Models Structured by Age and Genotype

The *age composition, growth rate*, and *genotype frequency* of two populations structured by age and allelic traits beginning with different initial conditions will be identical when projected into the distant future if they are both subject to the same selection regime and fixed schedules of age-specific birth and death rates (Charlesworth, 1980). Delays in attainment of genotypic frequencies in populations are caused by age structure effects and reproductive lags owing to the necessary inclusion of development times from newborn to mature adult.

Hierarchial Population Model (Honeybees)

Social insect populations such as the honeybee populations structured by individual age within a colony as well as by age of colony that are subject to a fixed regime of worker survival rates, queen reproduction, swarming thresholds, swarming properties, and colony survival rates with eventually attain a *stable individual age-by-colony distribution, a stable distribution of colonies by age*, and a *constant rate of colony increase*. The resulting population will consist of a fixed overall age distribution of individual worker bees within a fixed distribution of different aged colonies (S. D. Tuljapurkar, J. R. Carey, R. E. Page. A demographic theory for social insects: From individuals and colonies to populations, unpublished manuscript, 1992).

Models of Kinship

Individuals in two populations subject to the same fixed age-specific birth and death rates and projected into the distant future will, at specified ages, possess identical numbers of *descendents*, identical probabilities of having *living progenitors*, and equivalent numbers of *collateral kin* such as sisters, aunts, and nieces (Goodman et al., 1974; Pullum, 1982; Carey and Krainacker, 1988; Coresh and Goldman, 1988).

Parity Projection Models

If populations are structured not by age but by *parity*—the number of offspring individuals have produced—and then subject to parity-specific rates of death and future births, they will eventually attain a stable *parity distribution* and *rate of increase* that is independent of initial conditions (Feeney, 1983).

There are several reasons why understanding these generic properties of demographic models is important. *First*, only by understanding the general properties of models is it possible to begin to build general theories of population. For example, analogues of the intrinsic rate of increase and the stable age distribution are properties of all demographic models that can be framed in matrix form and possess ergodic properties. In addition, it is possible to address questions of the common characteristics of the modeled population, including growth rates, structure, momentum, convergence rates, sensitivity to parameter changes, and so forth. *Second*, a powerful set of mathematical and statistical tools can be brought to bear on all models. These tools do not have to be rediscovered for each new model. *Third*, understanding the interconnections among the demographic models fosters new perspectives and applications. For example, knowledge of stochastic models may suggest ways to incorporate stochasticity into multiregional models and, in turn, anticipate the "hybrid" properties of stochastic and multiregional models. *Fourth*, data available for one model might be directly applicable to another, thus reducing the need for a complete new set of expensive data.

One of the problems with all models, demographic ones notwithstanding, is that they are too often considered simply as predictors, and any inability to predict accurately is accepted as evidence of their uselessness. Stable population models were not constructed with the express intent of forecasting. As Gill (1986) notes, the demand that a scientific theory predict has been a stumbling block to serious study of process. She feels that the task of scientific theory is clear: it must explain; it may predict. For example, the phases of the moon can be predicted without understanding how it rotates about the earth or reflects the light of the sun. Similar arguments apply to the explanation and prediction of trends in natural populations.

BIBLIOGRAPHY

Athreya, K. B., and S. Karlin (1971) On branching processes with random environments. I. Extinction probabilities. *Ann. Math. Stat.* 42:1499–1520.

Bartlett, M. S. (1955) *An Introduction to Stochastic Processes with Special Reference to Methods and Applications.* Cambridge University Press, Cambridge.

Boyce, M. S. (1977) Population growth with stochastic fluctuations in the life table. *Theor. Pop. Biol.* 12:336–373.

Brian, M. V. (1965) *Social Insect Populations.* Academic Press, London.

Brian, M. V. (1983) *Social Insects. Ecology and Behavioural Biology.* Chapman and Hall, London.

Carey, J. R., and D. Krainacker (1988) Demographic analysis of tetranychid spider mite populations: extensions of stable theory. *Experimental and Applied Acarology* 4:191–210.

Caswell, H. (1989) *Matrix Population Models: Construction, Analysis, and Interpretation.* Sinauer Associates, Inc., Sunderland, Massachusetts.

Caswell, H., and P. A. Werner (1978) Transient behavior and life history analysis of teasel (Dipsacus sylvestris Huds.). *Ecology* 59:53–66.

Charlesworth, B. (1980) *Evolution in Age-structured Populations.* Cambridge University Press, Cambridge.

Charnov, E. (1982) *The Theory of Sex Allocation.* Princeton University Press, Princeton.

Cohen, J. E. (1979) Comparative statics and stochastic dynamics of age-structured populations. *Theor. Popul. Biol.* 16:159–171.

Coresh, J., and N. Goldman (1988) The effect of variability in the fertility schedule on numbers of kin. *Math. Popul Stud.* 1:137–156.

Feeney, G. (1973) Two models for multiregional population dynamics. *Environ. Planning* 5:31–43.

Feeney, G. (1983) Population dynamics based on birth intervals and parity progression. *Popul. Stud.* 37:75–89.

Frazer, B. D. (1972) Population dynamics and recognition of biotypes in the pea aphid (Homoptera: Aphididae). *Can. Entomol.* 104:1717–122.

Gill, S. P. (1986) The paradox of prediction. *Daedalus* 115 (3):17–48.

Goodman, L. A. (1953) Population growth of the sexes. *Biometrics* 9:212–225.

Goodman, L. A., W. Keyfitz, and T. W. Pullum (1974) Family formation and the frequency of various kinship relations. *Theor. Popul. Biol.* 5:1–27.

Hamilton, A. H. (1984). Sex ratio in the two spotted spider mite, *Tetranychus urticae* Koch: Theoretical and empirical investigations. Masters thesis, University of California, Davis.

Hamilton, A., L. W. Botsford, and J. R. Carey (1986) Demographic examination of sex ratio in the twospotted spider mite, Tetranychus urticae. *Entomol. Exp. Appl.* 41:147–151.

Hamilton, W. D. (1967) Extraordinary sex ratios. *Science* 156:477–488.

Harris, J. L. (1985) A model of honeybee colony population dynamics, *J. Apicul. Res.* 24:228–236.

Karmel, P. H. (1948) An analysis of the sources and magnitudes of inconsistencies between male and female net reproduction rates in actual populations. *Popul. Stud.* 2:240–273.

Kendall, D. G. (1949) Stochastic processes and population growth. *J. R. Stat. Soc. B* 11:230–264.

Keyfitz, N. (1980) Do cities grow by natural increase or by migration? *Geographical Analysis* 12:142–156.

Keyfitz, Nathan and John A. Beekman (1984) *Demography Through Problems*. Springer-Verlag, New York.

Kim, Y. J., and D. M. Strobino (1984) Decomposition of the difference between two rates with hierarchial factors. *Demography* 21:361–373.

King, B. H. (1987) Offspring sex ratios in parasitoid wasps. *Q. Rev. Biol.* 62:367–396.

Land, K. C., and A. Rogers (eds.) (1982) *Multidimensional Mathematical Demography*. Academic Press, New York.

May, Robert M (1973) *Stability and Complexity in Model Ecosystems*. Princeton University Press, Princeton.

McKendrick, A. G. (1926) Applications of mathematics to medical problems. *Proc. Edinburgh Math. Soc.* 40:98–130.

Michener, C. D. (1974) *The Social Behavior of Bees*. Harvard University Press, Cambridge.

Namboodiri, K., and C. M. Suchindran (1987) *Life Table Techniques and Their Applications*. Academic Press, Inc., Orlando, Fl.

Oster, G. R., and E. O. Wilson (1978) *Caste and Ecology in the Social Insects*. Princeton University Press, Princeton.

Pollak, R. A. (1986) A reformulation of the two-sex problem. *Demography* 23:247–259.

Pollak, R. A. (1987) The two-sex problem with persistent unions: A generalization of the birth matrix-mating rule model. *Theor. Popul. Biol.* 32:176–187.

Pollard, J. H. (1967) Hierarchical population models with Poisson recruitment. *J. Appl. Prob.* 4:209–213.

Pollard, J. H. (1969) A discrete-time two-sex age-specific stochastic population program incorporating marriage. *Demography* 6:185–221.

Pollard, J. H. (1973) *Mathematical Models for the Growth of Human Populations*. Cambridge University Press, Cambridge.

Pressat, R. (1985) *The Dictionary of Demography*. Bell and Bain, Ltd., Glasgow.

Pullum. T. W. (1982) The eventual frequencies of kin in a stable population. *Demography* 19:549–565.

Ryder, N. B, (1973) Two cheers for ZPG. *Daedalus* 102:45–62.

Ryder, N. B. (1975) Notes on stationary Populations. *Popul. Index* 41:3–28.

Rogers, A. (1965) The multiregional matrix growth operator and the stable interregional age structure. *Demography* 3:537–544.

Rogers, A. (1968) *Matrix Analysis of Interregional Population Growth and Distribution*. University of California Press, Berkeley.

Rogers, A. (1975) *Introduction of Multiregional Mathematical Demography*. John Wiley & Sons, New York.

Rogers, A. (1985) *Regional Population Projection Models*. Sage Publications, Beverly Hills.

Sakagami, S. F., and H. Fukuda (1968) Life tables for worker honeybee. *Res. Popul. Ecol.* 10:127–139.

Seal, H. L. (1945) The mathematics of a population composed of k stationary strata each recruited from the stratum below and supported at the lowest level by a uniform annual number of entrants. *Biometrika* 33:226–230.

Seeley, T. D. L. (1985) *Honeybee Ecology*. Princeton University Press, Princeton.

Tschinkel, W. (1988) Social control of egg-laying rate in queens of the fire ant, *Solenopsis invicta*. *Physiol. Entomol.* 13:327–000.

Tuljapurkar, S. D. and S. H. Orzack. (1980) Population dynamics in variable environments. I. Long-run growth rates and extinction. *Theor. Pop. Biol.* 18:314–342.

Tuljapurkar, S. D. (1982a) Population dynamics in variable environments. II. Correlated environments, sensitivity analysis and dynamics. *Theor. Pop. Biol.* 21:114–140.

Tuljapurkar, S. D. (1982b) Population dynamics in variable environments. IV. Weak ergodicity in the Lotka equation. *J. Math. Biol.* 14:221–230.

Tuljapurkar, S. D. (1984) Demography in stochastic environments. I. Exact distributions of age structure. *J. Math. Biol.*, 19:335–350.

Tuljapurkar, S. D. (1986) Demography in stochastic environments. II. Growth and convergence rates. *J. Math. Biol.*, 24:569–581.

Tuljapurkar, S. (1987a) Cycles in nonlinear age-structured models. I. Renewal equations. *Theor. Popul. Biol.* 32:26–41.

Tuljapurkar, S. D. (1987b) Population dynamics in variable environments. VI. Cyclical environments. *Theor. Popul. Biol.* 28:1–17.

Tuljapurkar, S. D. (1990) Population dynamics in variable environments. In: *Lecture Notes in Biomathematics, Vol. 85.* S. Levin (ed.). Springer-Verlag, Berlin.

Tuljapurkar, S., J. R. Carey and R. Page (1992) A demographic theory for social insects: From individuals and colonies to populations, unpublished manuscript.

Vajda, S. (1947) The stratified semistationary population. *Biometrika* 34:243–254.

Werner, P. A., and H. Caswell (1977) Population growth rates and age versus stage-distribution models for teasel (Dipsacus sylvestris Huds.). *Ecology* 58:1103–1111.

Wilson, E. O. (1971) *The Insect Societies*. Belknap Press of Harvard University Press, Cambridge.

Wilson, E. O. (1977) *Sociobiology*. Harvard University Press, Cambridge.

Wilson, E. O. (1985) The sociogenesis of insect colonies. *Science* 228, 1489–1495.

Winston, M. L. (1979) Intra-colony demography and reproductive rate of the Africanized honeybee in South America. *Behav. Ecol. Sociobiol.* 4:279–292.

Winston, M. L. (1987) *The Biology of the Honeybee*, Harvard University Press, Cambridge.

Young, A., and G. Almond (1961) Predicting distributions of staff. *Computer J.* 3:246–250.

Yule, G. U. (1924) A mathematical theory of evolution based on conclusions of Dr. J. C. Willis, F. R. S. *Phil. Trans. Royal Society*, B, 213, 21–87.

6
Demographic Applications

The purpose of this chapter is to fourfold. *First*, to apply some of the basic demographic principles to specific problems in population analysis. For example, I show how stable theory can be used to estimate growth rate in field populations. *Second*, to interrelate some of the standard tools in applied biology (e.g., probit analysis) to conventional demography (e.g., life table). *Third*, to show four different approaches to fitting curves that are common in biology and demography. *Fourth*, to demonstrate applications of demography to practical problems of insect mass rearing.

ESTIMATION

Growth Rate

The underlying concept for estimating population growth rate from age structure is best illustrated with the following example. Suppose the counts given in Table 6-1 were observed in a hypothetical population over three time steps with three age classes and that no mortality was experienced until the last possible age (age class 2).

Note that the ratio of any adjacent age class at any time is 2:1. For example, there is a twofold difference between the number in age class 0 and 1 as well as between the number in age class 1 and 2. These differences in numbers are completely due to the effect of growth rate on age structure. Therefore if individuals in the population experience no mortality until the end of their lives, we would know from the age relations that the population is growing daily by a factor of 2.

This relationship is made explicit by letting n_x and n_y denote the number in age classes x and y, respectively, c_x and c_y, the fraction of the total population in age classes x and y, respectively; and N, the total number in the population (after Keyfitz, 1985). Then the fraction of the population aged x is the number aged x divided by the total number in the population. Likewise the fraction aged y is the number aged y divided by the total number. That is:

$$c_x = n_x/N \qquad (6\text{-}1a)$$

$$c_y = n_y/N \qquad (6\text{-}1b)$$

DEMOGRAPHIC APPLICATIONS

Table 6-1. Hypothetical Age Structured Population Experiencing Complete Survival Between Age Classes That Is Increasing by two-fold Each time Step[1]

Time	0	1	2
Age 0	12	24	48
Age 1	6	12	24
Age 2	3	6	12
Total	21	42	84

[1] Note that the ratio of the number at age x to the number at age x + 1 at time t is 2 as is the ratio of the total number at time t + 1 to the total number at time t. These relationships illustrate the fundamental properties of stable populations—fixed growth rate and constant fraction of the total population in each age class.

and

$$c_x/c_y = n_x/n_y \tag{6-1c}$$

The point here is that the ratio of fractions in each of two age classes is equal to the ratio of the numbers.

The formulae for the fractions of the total stable population at ages x and y are

$$c_x = be^{-rx}l_x \tag{6-2a}$$

$$c_y = be^{-ry}l_y \tag{6-2b}$$

where r denotes the intrinsic rate of increase, b denotes the intrinsic birth rate, and l_x and l_y denote survival to ages x and y, respectively. Therefore the ratio of c_x to c_y is

$$\begin{aligned} c_x/c_y &= (be^{-rx}l_x)/(be^{-ry}l_y) \\ &= (e^{-rx}l_x)/(e^{-ry}l_y) \end{aligned} \tag{6.3}$$

Since $c_x/c_y = n_x/n_y$, then

$$\begin{aligned} n_x/n_y &= (e^{-rx}l_x)/(e^{-ry}l_y) \\ &= \{e^{-r(y-x)}l_x\}/l_y \end{aligned} \tag{6-4}$$

Rearranging and taking logs yields a solution for r as

$$r = \frac{1}{(y-x)} \ln_e \left[\frac{n_x l_y}{n_y l_x} \right] \tag{6-5}$$

An example of the application of this technique is given in Table 6-2 for mite-infested plants, where the mean age of the mite egg stage is 2.5 days ($= x$), the mean age of the immature stage is 7.5 days ($= y$), and no mortality occurs between stages. Therefore $(y - x) = 5$ days, $l_x = l_y = 1.0$. Growth rate of mites on plant 1 is computed as $r = (1/5)[\ln(2311/824)] = .206$ or $\lambda = e^r = 1.23$.

All of these different growth rates are within the range possible in spider mites (Carey, 1982) and suggest that some populations are increasing while

Table 6-2. Number of Spider Mite Eggs and Immatures and Estimate of Population Growth Rate, r, based on Equation 6-5 (Data from Carey, 1983).

	Plant Number				
	1	2	3	4	5
Eggs	2311	1150	2048	3654	2145
Immatures	824	596	541	5820	4643
Estimated r	.206	.131	.266	−.093	−.154

others are decreasing. The finite rates of increase for the mite populations on plants 1 throught 5, respectively, are 1.23, 1.14, 1.30, 0.91, and .86, with a mean geometric growth rate of $\lambda = 1.07$. Therefore the average mite population on each of the plants increased daily by a factor of 1.07.

Insect-Days

Insect-days represent the sum of products of the number of individuals and time. This concept is useful in basically two contexts. *First*, it allows the different periods lived by individuals to be taken into account proportionately when calculating rates. For example, in the life table the days lived by individuals that survive to old age carry the same weighting as the days lived by individuals that die at younger ages. This also holds for days lived within an interval used for computing L_x in the life table. *Second*, it provides a conceptual foundation for normalizing population duration and numerical weightings in ecology and pest management. Two questions are commonly addressed using the insect-day concept: i) how many insect-days were lived during an intercensal period? This provides a basis for finding the average population number over a given period; and ii) how many individuals lived in this intercensal period. This is possible if an estimate of the expectation of life for individuals in the population can be made. The advantage here is that insect-days provide a common unit for comparison between two populations in terms of both insect-days and in lifetimes.

Consider a population with 200 individuals at a first census and 500 at a second census 3 days later. In this case the population grew by 2.5-fold in 3 days or an average of 1.357-fold per day (i.e., $\sqrt[3]{2.5}$). The number of insect-days that were lived in this interval can be approximated in several ways, two of which are given below:

1. *Trapezoidal method*. Here the population curve is viewed as a straight line between 200 and 500, which then forms the sides of a trapezoid of length 3 days. Thus to obtain the number of insect-days under this curve we can use the formula for the area of a trapezoid. i.e., the sum of the "bases" times the "length" divided by 2:

Table 6-3. Computation of Mite-Days per Cotton Plant over a Gowing Season in the San Joaquin Valley, California[1]

Time (t)	Population at t	Midpoint Number
0	200	
		236
1	271	
		320
2	320	
		434
3	500	
	Total	990 insect-days

$$\text{insect-days} = t_3 \times (N_0 + N_3)/2$$
$$= 3 \times (200 + 500)/2$$
$$= 1050$$

Therefore the population average was estimated at 1050/3, or 350, on each of the 3 days.

2. *Geometric summation.* The assumption for this case is geometric change between census times. This approach can be illustrated by using the average daily growth rate to "project" the population from t_0 to t_3 as follows, starting at $t = 0$ with 200 individuals in the populations increasing daily by a factor of 1.357.

Therefore the estimated number of insect days is 990 or an average of 330 insects each day.

A more accurate method for estimating the area under a geometric growth curve is to use the formula for the sum of a finite geometric series.

$$N_0\lambda^0 + N_0\lambda^1 + N_0\lambda^2$$
$$= \sum_{i=0}^{k} N_0 \lambda^{i-1}$$
$$= N_0(\lambda^3 - 1)/(\lambda - 1) \quad (6\text{-}6)$$

where

$$\lambda = \sqrt[3]{N_3/N_0}$$

This formula is the discrete version of the integral of population number from times t_0 to t_3. For this case $N_0 = 200$, $\lambda = 1.357$, thus

$$\text{insect-days} = N_0(\lambda^3 - 1)/(\lambda - 1)$$
$$= 200(2.5 - 1)/(1.357 - 1)$$
$$= 200(1.5)/.357$$
$$= 840$$

For this case the population average was estimated as 840/3, or 280, each day.

Table 6-4. Computation of Mite-Days per Cotton Plant Over a Growing Season in the San Joaquin Valley, California[1]

Date	Week	Number per Plant	X-Fold Change Interval	X-Fold Change Daily	Mite-Days
April 15	−10	1.0			
			50.3	1.06	856.4
June 28	0	50.3			
			3.38	1.19	630.7
July 5	1	169.8			
			76	.96	1,018.8
July 12	2	128.8			
			3.73	1.21	1,674.4
July 19	3	480.0			
			4.32	1.23	6,928.7
July 26	4	2,072.1			
			1.37	1.05	15,333.5
Aug. 2	5	2,839.8			
			3.59	1.20	36,775.4
Aug. 9	6	10,190.3			
			.93	.99	66,404.3
Aug. 16	7	9,448.3			
			.88	.98	63,842.8
Aug. 23	8	8,334.5			
			.58	.93	47,414.9
Aug. 30	9	4,872.2			
			.06	.82	13,878.4
Sept. 13	11	303.0			
				Total	254,758.3

[1] Planting date was April 15. A total of 856.4 mite-days were estimated to have accumulated from April 15 to June 28, when the first sample was taken. This was based on arbitrary infestation rate of one mite/plant at the start of the season (data from Carey, 1983).

An example of the application of this method to data on spider mite populations on cotton plants is given in table 6-4 and Figure 6-1. As an example computation, the number of mite-days lived in the interval from week 2 to week 3 is computed in two steps as follows:

Step 1. Compute the daily growth rate by first finding the factor by which the population increased over the 7-day period and then taking its 7th root.:

$$N_3/N_2 = 480/129$$
$$= 3.72\text{-fold change in 7 days.}$$

The daily change λ is then $\sqrt[7]{3.72} = 1.21$-fold per day.

Step 2. Use the geometric formula for sum of finite series to determine insect-days:

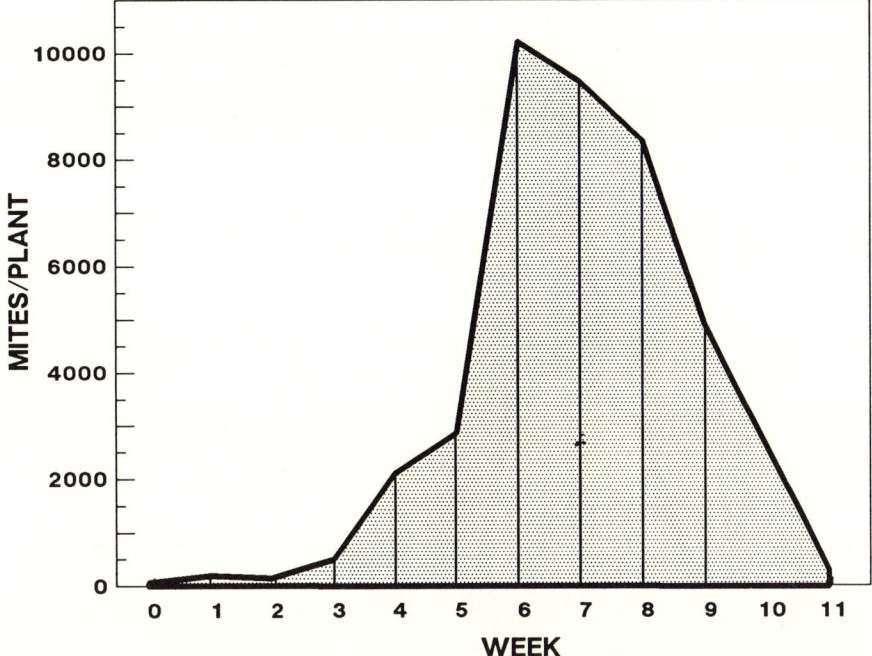

Figure 6-1. Illustration of population area curves for determining insect (i.e., mite) days over a growing season (data from Carey, 1982).

$$\text{insect-days} = N_2 \left[\frac{(N_3/N_2) - 1}{\lambda - 1} \right]$$

$$= 129 \left[\frac{3.72 - 1}{1.21 - 1} \right]$$

$$= 1674$$

The sum of insect-days over each of the 11 intervals is 254,758 mite-days per plant. Assuming that the average mite lived 14 days, then the total number of mite-days divided by the average life expectancy yields an approximation to the average number of mites that ever lived on a single plant through the season. This number is 18,197 mites that ever lived per plant. In other words, one cotton plant provided the necessary resources for the production of over 18,000 mites.

Stage Duration

A problem that has received considerable attention in ecology and entomology is determining the time required for insects to attain various stages in their development and survival to those stages. The approach is often referred to as the analysis of stage-frequency data (e.g., Bellows and Birley, 1981;

Table 6-5. Two-spotted Spider Mite Development (Carey, unpublished)

Time (1)	Eggs (2)	Immatures (3)	Adults (4)	Total (5)
0	100			100
1	100			100
2	100			100
3	87	12		99
4	5	90		95
5		87		87
6		82		82
7		79		79
8		60		60
9		42	15	57
10		30	26	56
11		19	35	54
12		2	49	51
13			50	50

Mills, 1981; Braner and Hairston 1988; Manly and McDonald 1989; Pontius et al. 1989).

The general problem can be described with the following example. Suppose 100 newly laid mite eggs were placed on a single leaf. Each day the number in each stage (i.e., egg, immature, adult) is counted until all individuals have either completed development or died. Because considerable developmental variation exists, more then one of the stages is present on some sampling dates. Data for mite development are presented in Tables 6-5 and 6-6.

Table 6-6. Life Table for Estimating Development Duration from Developmental Data on Mites from Table 6-5

	Survival		Egg Development			Egg + Immature Development		
x or t	l_x	d_x	l_t	L_t	$d_x(e_0-t)^2$	l_t	L_t	$d_x(e_0-t)^2$
0	1.00	.00	1.00	1.00	.00	1.00	1.00	.00
1	1.00	.00	1.00	1.00	.00	1.00	1.00	.00
2	1.00	.01	1.00	.94	.02	1.00	1.00	.00
3	.99	.04	.87	.46	.28	1.00	1.00	.00
4	.95	.08	.05	.03	.12	1.00	1.00	.00
5	.87	.00	.00	.00	.00	1.00	1.00	.00
6	.87	.08	.00	.00	.00	1.00	1.00	.00
7	.79	.19	.00	.00	.00	1.00	1.00	.00
8	.60	.03	.00	.00	.00	1.00	.87	.36
9	.57	.01	.00	.00	.00	.74	.64	.01
10	.56	.02	.00	.00	.00	.54	.45	.13
11	.54	.03	.00	.00	.00	.35	.19	1.04
12	.51	.01	.00	.00	.00	.04	.02	.32
13	.50	—	.00	.00	.00	.00	.00	.00
		.50		$e_0=3.42$	.42		$e_0=10.17$	1.85

The objective is to determine the average duration of each of the first two stages. This can be accomplished using a life table approach, where the exit from one stage to another is similar to dying (i.e. alive to dead states). There are three ways to exit an age class: i) by surviving to the next age within a stage; ii) by surviving to the next age and into the next stage; and iii) by dying. Each individual is subject to two competing risks: the risk owing to the "force of mortality" and the risk owing to the "force of transition". Although some authors have suggested that it is possible to obtain stage-specific development *and* survival from the data (e.g., Bellows and Birley, 1981; Manly, 1985), it is not clear that this is logically or even theoretically possible (see Hairston and Twombly, 1985). The reason for this can be illustrated with the following example. Consider the subset of the data from Table 6-5 presented in Table 6-7.

Note the following: i) there were four individuals that died in the age interval 3 to 4. The origin of these four individuals is the central issue: ii) the number of immatures increased from 12 at age 3 to 90 at age 4. The difficulty arises at this point because it is not possible to distinguish the "force of mortality" from the "force of transition" without knowing which individuals died. That is, we cannot tell whether the four that died were all eggs, all immatures, or some fraction of the original number at risk in age 3. The point here is that stage-specific survival is not possible to obtain directly from these types of data although apparently reasonable approximations can be obtained using various assumptions about distributions and numerical methods (e.g. Braner and Hairston, 1988).

The two sets of parameters that can be obtained directly from these data are i) means and variances of developmental times through each stage; and ii) age-specific but not stage-specific survival. A framework for obtaining this information is presented in Table 6-6. Note that age-specific survival data are based on total numbers surviving to each time period in Table 6-5.

Tabulations needed to obtain development times are given in columns 4 through 9. The concept involves the use of the life table, where l_t and L_t are the analogues of the l_x and L_x life table schedules, respectively. Each is based on the fraction of those individuals that ultimately attain a stage that *have not* attained the stage by time t. For example, the "transition schedule" from the egg stage is simply the column corresponding to the number in this stage

Table 6-7. Subset of Spider Mite Developmental Data from Table 6-5 Showing the Number of Eggs and Immatures at Age 3 and 4 days (see Text)

Age	Number		
	Eggs	Immatures	Total
3	87	12	99
4	5	90	95

at time t in Table 6-5. Example computations for the egg + immature development times are:

$l_7 = 79$ (not attained adulthood)/79 (at risk)
$= 1.00$
$l_9 = 42$ (not attained adulthood)/57 (at risk)
$= .74$

Once the l_t schedules are constructed for the duration within a stage, the L_t schedule and sums of squares columns can be tabulated to obtain mean duration and variances. The sum of the L_t columns will give the respective "expectation of duration at birth". These are

egg stage duration = 3.42 days
egg stage developmental variance = 0.42
egg + immature duration = 10.17 days
egg + immature developmental variance = 1.85
immature stage duration = 10.17 − 3.42 = 6.75 days
immature stage developmental variance = 1.85 − .42 = 1.43

The underlying reasoning for this approach was made explicit by Pontius et al. (1989), though they presented their method in a statistical and not a demographic context.

Mosquito Survival Analysis using Parity Rates

Suppose we wish to estimate daily survival in a field population of mosquitoes. The basic concept can be understood by considering a newly emerged cohort of female mosquitoes subject to a daily survival rate of 0.90 whose gonotrophic cycle length is 5 days. The lifetime survival schedule for this cohort will be a geometrically decreasing function of age (Figure 6-2) with an expectation of life of 10 days lived by the average newly emerged female. These days-lived can be partitioned into i) nulliparous-days, which equal 4.1 days; and ii) parous-days, which equal 5.9 days. Therefore for this example the percentage of all mosquito-days lived by the cohort in the parous state is 59% (i.e., 5.9 days/10.0 days).

Two aspects of this model will alter the percentage of parous-days i) daily survival will affect the slope of the curve and therefore the age distribution; and ii) gonotrophic cycle length will change the number of days lived in the nulliparous period. For example, a daily survival rate of 0.80 will produce a steeper schedule than the one in Figure 6-2, and therefore a lower fraction of all mosquito-days will occur in the parous state. Likewise an increase in the length of the gonotrophic cycle from 5 to 6 days will increase the number of days in the nulliparous state by one and therefore decrease the percentage of all days lived in the parous state. The concept for application of this model to field data is that only one combination of daily survival and gonotrophic cycle length will yield a given percent parous-days. Therefore if any two of the parameters are known, the third can be easily determined. More

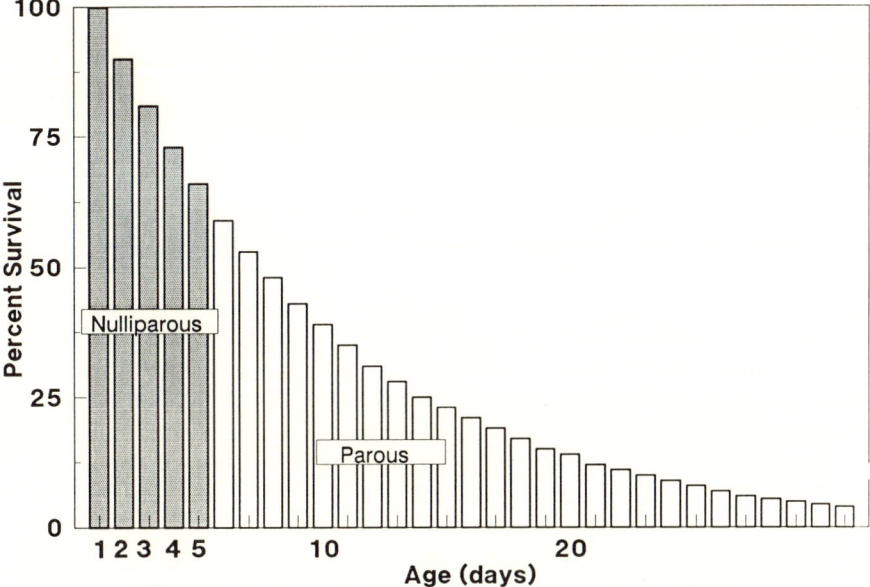

Figure 6-2. Age distribution and partitioning of mosquito-days in a cohort subject to a daily survival of .90 and with a 5-day gonotrophic cycle length. The demographic model is based on the assumption that a particular distribution of nulliparous- and parous-days can be generated only by a particular combination of daily survival and gonotrophic cycle length (GCL). Therefore if both GCL and parity rate (percent parous-days) are known, daily survival can be computed.

specifically, if gonotrophic cycle length and parity rate are known, daily survival can be computed.

An important aspect of the application of this model is the correct interpretation of field parity rates. A seasonal parity rate of 30% implies that the average mosquito that ever lived spent 30% of its life in the parous state. Parity is a measure of population structure that is determined, in part, by the particular survival regime to which a population is subjected. Another important factor that affects population structure is growth rate. Therefore it is invalid to apply this model to data on weekly or monthly parity rates because increasing or decreasing populations at the time of sampling will shift the age structure away from the stationary case (i.e., zero population growth). It is designed for application to seasonal parity rates since the increases will offset the decreases over a season.

Davidson (1954) showed that daily survival equals the nth root of the parity rate in a stationary population where n is the length of the gonotrophic cycle. Note that the 5th root of the seasonal parity rate of 0.59 in the example is 0.90, or the daily survival rate. This is the main result in that parity data yields three pieces of information: i) daily survival ($=90\%$); ii) fraction of all mosquito-days in the parous state ($= 0.59$); and iii) fraction of all newly emerged mosquitoes that survivied to the parous state ($= .59$). Thus seasonal parity has a threefold interpretation.

Pesticide Effective Kill Rate

A problem encountered in determining the effectiveness (e.g., kill or sterility rate) of a chemical, radiation, or temperature treatment in insect bioassay is separating out treatment-induced effects from effects that would have occurred in the absence of the treatment. For example, a typical insecticide bioassay design involved replicates of cohorts subjected to the test chemical as well as replicates of cohorts not subjected to the chemical (i.e., check or control). After a predetermined period (i.e., 24–48 hrs), the proportion that are dead in each of the two sets of cohorts is recorded. The arithmetic problem is concerned with adjusting for the natural mortality that may have occurred in the treatment in order to isolate the "pure" effectiveness of the chemical (Carey, 1986).

Abbot (1925) derived the original formula for natural mortality adjustment in bioassay by introducing the concept of an "effective kill rate" (EKR) of a chemical. This formula is (notation different than Abbot's):

$$EKR = (P_{nat} - P_{trt})/P_{nat} \qquad (6\text{-}7)$$

where P_{nat} denotes the proportion alive in the check, P_{trt} denotes the proportion alive in the treatment, and $(P_{nat} - P_{trt})$ gives the proportion dead owing to the treatment alone. This adjustment is known as Abbot's correction in toxicology and is accepted as convention. It should be noted that the implicit assumption in determining EKR is that the two probabilities of dying in the treatment are independent.

My purpose for addressing Abbot's correction is to demonstrate that the concept is a multiple (double) decrement life table over a single time (age) interval. The major conceptual difference between Abbot's approach and the demographic approach is that of derivation perspective—Abbot's concerns mortality probabilities, and the demographic focuses on survival probabilities.

The starting point for the re-derivation of Abbot's EKR is to view the probability of surviving a treatment (P_{trt}) as conditional on the product of the probability of surviving natural mortality (P_{nat}) and the probability of surviving the chemically induced mortality (P_{chem}). These correspond to i causes (i = 1, 2) in multiple decrement theory and are competing risks. Thus

$$P_{trt} = (P_{nat})(P_{chem}) \qquad (6\text{-}8a)$$

The proportion remaining alive in the treatment plot or cage equals P_{trt} and the proportion remaining alive in the check equals P_{nat}. Therefore, the probability of surviving the chemical in the absence of natural mortality is found by rearranging Equation 6-8a as

$$P_{chem} = \frac{P_{trt}}{P_{nat}} \qquad (6\text{-}8b)$$

DEMOGRAPHIC APPLICATIONS

Since P_{chem} denotes the probability of surviving the chemical in the absence of natural mortality, its complement represents the probability of dying owing to the chemical alone. This corresponds to Abbot's EKR. Thus

$$EKR = 1 - P_{chem}$$

$$= 1 - \frac{P_{trt}}{P_{nat}}$$

$$= \frac{P_{nat} - P_{trt}}{P_{nat}} \qquad (6.8c)$$

Two points of this re-derivation merit comment. *First*, by shifting emphasis from mortality rate to survival rate the derivation becomes more mathematically transparent. *Second*, by establishing the parallel between Abbot's correction and multiple decrement life table analysis, the purpose of both becomes more apparent. Moreover, the multiple time and multiple treatment version of Abbot's correction becomes the multiple decrement life table.

Consider the following example. A cohort of 1000 melon flies was treated with .2% of the insecticide DDT; 76%, 71%, and 68% remained alive after 2, 5, and 15 days, respectively, while 96%, 93%, and 91% of the controls were alive after these respective periods (from Carey, 1986). Questions: i) What is the effective kill rate of DDT over the entire period? and ii) What are these rates over each successive period?

The first question can be answered by noting that $P_{trt} = .68$ and $P_{nat} = .91$. Therefore

$$P_{chem} = P_{trt}/P_{nat}$$
$$= .68/.91$$
$$= .75$$

Thus the effective kill rate (EKR) is

$$EKR = 1 - P_{chm}$$
$$= 1 - .75$$
$$= .25$$

Therefore the proportion of fruit flies killed due to the insecticide DDT alone (i.e., in the absence of natural mortality) is .25.

The second question is most easily addressed in the form of a multiple decrement table by computing the P_{chem} (Abbot's correction) and P_{nat} for each day. The results are shown in Table 6-8.

Therefore, the effective kill rate for the period 0–2 days is $(1 - .79) = .21$, for 3–5 days it is $(1 - .97) = .03$, and for 6–15 days it is $(1 - .98) = .02$. Additional insight into these data may be gained by converting these effective kill rates over these unequal time periods to average daily effective kill rates. This is done by taking the root of P_{chem} corresponding to the period of observation to obtain average daily survival probabilities and then substracting this

Table 6-8. Multiple Decrement Life Table Technique Applied to Chemical Insecticide Data for Melon Flies (from Carey, 1986)[1]

Day (x)	Probability		Period Survival, P_x	Total Survival, l_x
	P_{chem}	P_{nat}		
0	.79	.96	.76	1.00
2	.97	.97	.94	.76
5	.98	.98	.96	.71
15	—	—	—	.68

[1] The term P_{chem} is Abbott's Correction (Effective Kill Rate) and gives the probability of surviving chemical application in the absence of natural mortality, and p_{nat} is the probability of surviving in the absence of chemical. Period survival, P_x, is the probability of surviving both causes of death from x to x + 1, and total survival, l_x, is the probability of surviving to age x in the presence of both causes of death.

value from 1.0 to obtain daily mortality (kill) rates. For example, the square root (i.e., 2-day period) of $P_{chem}(=.79)$ for the 0–2-day period is .89 and its complement (daily effective kill rate) is .11; the cube root of $P_{chem}(=.97)$ for the 3–5-day period is .99 and its complement is .01; and the 9th root of $P_{chem}(=.98)$ for days 6–15 is .998 and its complement is .002. Hence the average daily EKR for the first 2 days after exposure (i.e., the residual) is around 10-fold greater than the daily EKR for the next 3 days and 50-fold greater than the daily EKR from 6 to 15 days after initial exposure.

Probit Analysis: Life Table Perspectives

The pioneer analyst of quantal response in bioassay was Finney (1964), who is best known for developing the technique known as probit analysis. The basic notion for this method can be described as follows. An assumption is made that mortality of organisms increases arithmetically as the dose of a toxicant to which they are subjected increases geometrically. Hence, by transforming arithmetic dose to $\log_e$ dose, the percent mortality per increment of $\log_e$ dose is normally distributed (approximately) and the proportion dying is equal to the area under the normal density curve below a given dose. This is referred to as a "probit" (= *prob*ability un*it*). The $\log_e$ dose at which 50% mortality occurs (e.g. LD_{50}) is set at probit 5. This dose plus 1 standard deviation is probit 6 (e.g., LD_{68}), plus 2 standard deviations is probit 7 (e.g. LD_{95}), and so forth. The determination of the LD_{50} originated as a measurement of potency of a drug whose chemical purity could not be assayed at the time. It was bound up with the idea, still persisting, that for each drug and species an absolute LD_{50} value existed (Morrison et al., 1968).

The statistics of the life table are identical with those of probit analysis if age is viewed as a dose of time. The connections between probit analysis and life table analysis can be established as follows:

1. Life table and probit analysis are both concerned with same quantal response (i.e., alive/dead).

2. The d_x schedule of the life table is the analog of the normalized distribution of deaths per increment of dose in probit analysis. For example, all deaths that occur in a series of bioassays covering a range of toxicant doses are normalized so that their sum equals unity (Litchfield and Wilcoxon, 1949). This establishes the frequency distribution of deaths in a hypothetical cohort.
3. The $\log_e$ dose vs. mortality curve in a bioassay is typically presented as the cumulative normal distribution that represents cumulative mortality with increasing dose. The l_x parameter of the life table is the complement of this curve. Thus the curve $(1 - l_x)$ vs. age is conceptually identical with the mortality vs. dose relationship in bioassay.
4. It is assumed in probit analysis that the $\log_e$ dose at which 50% mortality occurs (LD_{50}) represents both the mean dose and the median dose (Morrison et al., 1968). The expectation of life at birth, as noted earlier, can be redefined as the exact mean age of death at birth of the cohort and usually is close to the median age of death (age at which $l_x = .5$). Hence, zero dose corresponds with age zero (birth) in a cohort, and the LD_{50} of probit analysis corresponds exactly with the expectation of life at birth of the life table. Therefore, the parameter, e_0, can be reinterpreted as the lethal "dose" of time at which 50% of a cohort is dead.
5. The number of standard deviations added to e_0 in a life table will give the area under the normal curve and the corresponding expected level of cohort mortality. For example, $e_0 + 1$ SD represents 68% mortality, $e_0 + 2$ SD represents 95% mortality, and so forth. This is identical with probit analysis.
6. The slope is log dose-response curve will indicate the relationship between the change of dose and the lethal response. This slope is felt to be more important in risk assessment than the LD_{50} because more insight about the intrinsic toxic characteristics of a compound is available. Sometimes the slope suggests the mechanism of toxicity. For example, a steep slope may indicate rapid onset of action or faster absorption. This corresponds with the "force of mortality," or the q_x schedule in the life table.

The demographic ties between probit analysis and the life table may illustrated with the following example. Back and Pemberton (1916) examined the effect of cold temperatures upon medfly eggs by refrigerating them within a host and removing subsamples on preselected days. The data were presented as the number of eggs remaining alive vs. the number under observation for the given period. For example, egg trials at 0°C for the first 4 of 15 days revealed that all of the 81 eggs hatched that were observed for 1-day exposure, 520 out of 528 eggs hatched after a 2-day exposure, 135 out of 150 hatched after a 3-day exposure, and 216 out of 336 eggs hatched after a 4-day exposure. Questions: i) How would one construct a life table from these data? ii) What are probits 5-9? iii) What are the 95% confidence limits for the proportion dead after 4 days of cold exposure?

For the first question, these data simply give the probability of an individual's surviving to day x at 0°C. Therefore, $l_0 = 1.00$ (by definition), $l_1 = 1.00$ ($= 81/81$), $l_2 = .98$ ($= 520/528$), $l_3 = .90$ ($= 135/150$), $l_4 = .64$ ($= 216/336$), and so forth. The d_x schedule can then be constructed from this l_x curve and the mean and standard deviation computed.

For the second question, the expectation of life and life table variance from Back and Pemberton's data were 3.61 days and .71, respectively. The results of adding 1 standard deviation for every probit over probit 5 are shown in Table 6-9.

For the third question, the proportion of the 336 eggs that survived the 4-day cold exposure was 0.64 ($= 14$), thus the proportion that died was .36. The formula for a sample variance is

$$S^2 = q(1-q)/n$$

where q is the probability of dying ($= .36$) and n is the number observed ($= 336$). Therefore the variance and standard deviation for these egg data are

$$S^2 = (.36)(.64)/336$$
$$= 0.00068$$
$$S = .0261$$

The 95% confidence intervals for the proportion dying after the 4-day exposure are

$$CI_{95} = q \pm 1.96S$$
$$= .34 \pm .05$$

Thus we conclude with 95% confidence that the probability of a medfly egg's dying after 4 days' exposure to 0°C is between .29 and .39.

Percent Parasitism

A standard measure in many field studies on insect parasitization is "percent parasitism." This usually involves collecting preadults (e.g., larvae), allowing them to mature (e.g., pupate), and recording the percentage of those that are parasitized out of all collected. The problem is to determine what this is actually measuring.

The simplest way to envision the problem is to consider the three

Table 6-9. Probit Analysis Applied to Lethal Temperature Data on Fruit Fly Eggs (Data from Back and Pemberton, 1916)

Probit	Days at 0°C	LD
5	3.61	LD_{50}
6	4.32	LD_{68}
7	5.03	LD_{95}
8	5.74	$LD_{99.9}$
9	6.45	$LD_{99.997}$

Table 6-10. Examples of Parasitism Rates for Three Hypothetical Cases Illustrating the Problem with Interpretation of Percent Parasitism if "Other" Causes of Death Are Unobservable[1]

Alive/Dead	Case		
	I	II	III
Alive	20	10	5
Dead			
Parasitized	40	20	10
Other	40	70	85

[1] In all three cases the ratio of parasitized to alive is 2 yet the proportion of total deaths due to parasitism ranges from 50% (Case I) to 12% (Case (III)).

hypothetical cases of parasitism (after Carey, 1991) given below, where there are two sources of mortality: i) larval parasitoids and ii) "other," which cannot be measured or observed (see Table 6-10).

Note in all three cases the observed fraction parasitism is .33. For example, case I is 20/(20 + 40); case II is 10/(10 + 20); and case III is 5/(5 + 10). So we could naively conclude that parasitism in all three cases is identical. Yet when we observe the "unobservable" (i.e., "other" causes), it is evident that half of all deaths are due to parasitism in case 1 (i.e., 40/80) but only slightly over 10% of all deaths in case III (i.e., 10/95). This type of problem may lie at the heart of the ambiguity of results in evaluating the outcome of many biological control programs. It is exceedingly difficult to obtain the estimates of total lifetime or preadult mortality that are needed to make statements regarding the *fraction of total mortality* attributable to parasitoids.

Life History Scaling

Two major problems exist when comparing insect life histories: i) temperature and other aspects of rearing usually differ between experiments, and ii) age scales may differ drastically. By expressing age in terms of one of several demographic parameters (e.g., expection of life), all other parameters can be scaled to this baseline and the reproductive schedule expressed as a percentage within each of the resulting intervals. A normalized life history schedule for the medfly is given in Figure 6-3. Note that i) half of all eggs are laid by the mean age of reproduction and 5/6 of all eggs laid by the expectation of life at eclosion; ii) only 1/6 of all eggs are laid by the cohort from the mean age of death to the last day of life (life span); and iii) mean age of reproduction is half the expectation of life and a quarter of the life span (Carey, 1991).

CURVE FITTING

Curve fitting is useful in demography for basically five reasons (modified from Keyfitz, 1982): i) *Smoothing data*—makes the data easier to handle and

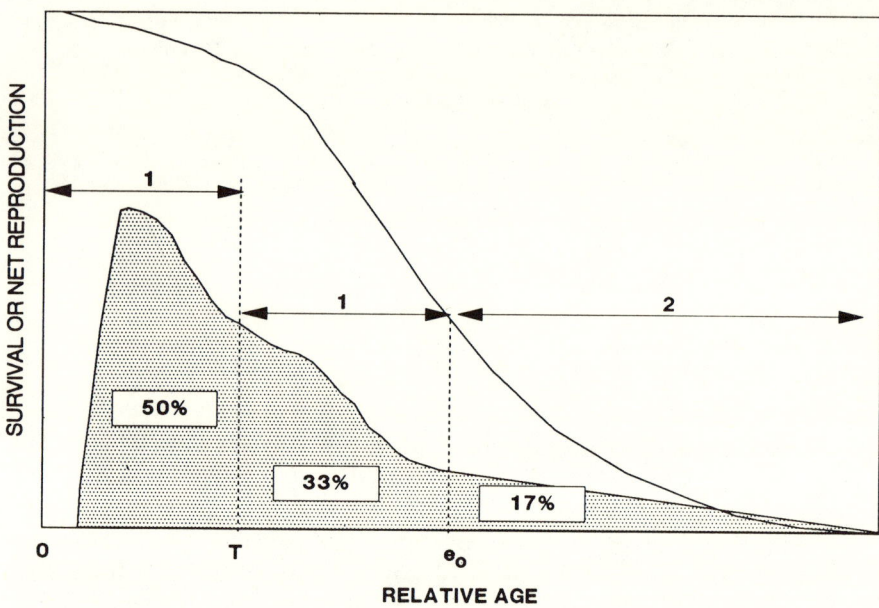

Figure 6-3. Normalized medfly life history where e_0 denotes the expectation of life at eclosion and T the mean age of net reproduction.

removes irregularities and inconsistencies. For example, a survival curve near the tail is often based on few individuals and therefore contains days in which no mortality occurs. ii) *Increase precision*—the assumption here is that the "real" pattern underlying the observations is a smooth curve. iii) *Aid inferences from incomplete data*—this use is basically interpolation when data are not available for intermediate points. For example, many temperature development studies use increments of 5–10°C rather than observe development under all temperatures. iv) *Facilitate comparisons*—many times a single parameter is sufficient to characterize differences or similarities between two sets of observations. For example, two curves fitted to data on logistic growth can be compared using the carrying capacity parameter, K. v) *Aid forecasting*—curve fits are used for parameters in population forecasting models as well as simple extrapolation from observed population trends.

Curve fitting generally involves two separate steps (Carnahan et al., 1968). *First*, the mathematical expression that relates one set of observed values to a second set is determined. This step is often the most difficult because it is usually judgmental. *Second*, finding the appropriate parameter values for a best fit. This set of parameter values is typically found through the technique known as "least squares" or by ad hoc methods such as parameter adjustment until the curve is fit visually. Four separate approaches will be shown—least squares (linear equation), solution for parameter values using simultaneous

DEMOGRAPHIC APPLICATIONS

equations (logistic equation), ad hoc parameter estimation (Gompertz), and visual fit (Pearson type I curve).

The Linear Equation: Least Squares

The concept of least squares curve fit for a linear equation involves i) establishing a separate function, S, for the sum of the squares of the departures of a linear equation from the observed data; and ii) applying the calculus of extrema to find the minimum of this function. We proceed by describing the function S, finding the derivatives of this function with respect to m (i.e., ∂m) and b (i.e., ∂b), setting these equal to zero, and solving for m and b in the linear equation $Y = mX + b$.

$$S = \sum(y_i - mX_i - b)^2 \qquad (6\text{-}9a)$$

$$\partial S/\partial m = -2\sum X_i(y_i - mX_i - b) \qquad (6\text{-}9b)$$

$$\partial S/\partial b = -2\sum(y_i - mX_i - b) \qquad (6\text{-}9c)$$

For S to be a minimum, $\partial S/\partial m$ and $\partial S/\partial b$ must equal zero, so

$$0 = \sum X_i(y_i - mX_i - b) \qquad (6\text{-}10a)$$

$$0 = \sum(y_i - mX_i - b) \qquad (6\text{-}10b)$$

Rearranging, we obtain the simultaneous normal equations

$$nb + m\sum X_i = \sum y_i \qquad (6\text{-}11a)$$

$$b\sum X_i + m\sum X_i^2 = \sum X_i y_i \qquad (6\text{-}11b)$$

where n is the number of observations. These equations can be solved for m and b as

$$m = \frac{n\sum X_i Y_i - \sum X_i \sum Y_i}{n\sum X_i^2 - (\sum X_i)^2} \qquad (6\text{-}12a)$$

$$b = \frac{\sum Y_i - m\sum X_i}{n} \qquad (6\text{-}12b)$$

As a numerical example, consider the data set for development rate of the melon fly at six different temperatures (Table 6-11).

Table 6-11. Developmental Rates for the Melon Fly at six Temperatures (Data from P. Yang, Unpublished)

	Trial (i)					
	1	2	3	4	5	6
Temperature (X_i)	19	22	25	28	30	32
Development (Y_i)	.040	.044	.056	.072	.082	.084

Thus

$$n = 6$$
$$\sum X_i^2 = (19)^2 + (22)^2 + (25)^2 + (28)^2 + (30)^2 + (32)^2 = 4178$$
$$\sum X_i = 19 + 22 + 25 + 28 + 30 + 32 = 156$$
$$\sum Y_i = (.040) + (.044) + (.056) + (.072) + (.082) + (.084) = .378$$
$$\sum X_i Y_i = 19(.040) + 22(.044) + 25(.056) + 28(.072) + 30(.082) + 32(.084)$$
$$= 10.292$$

and

$$m = \frac{6(10.292) - 156(.378)}{6(4178) - 156^2}$$
$$= .0038$$
$$b = \frac{.378 - (.0038)156}{6}$$
$$= -.0358$$

The best-fitting straight-line equation for this set of temperature development data is thus

$$Y = .0038X - .0358$$

For this particular application of developmental rate regressed on temperature, the inverse of the slope $m = .0038$ gives the number of degree-days required for development, which is 263 degree-days. The developmental zero is determined by setting $Y = 0$ and solving for X (= temperature), which is 9.4°C.

Population Curve: The Logistic Equation

Parameters for the logistic equation can be determined from only three population samples taken at equidistant times by formulating three equations in three unknowns. The solution for the logistic differential equation is set equal to the population levels at each of the three sample dates, which creates the three simultaneous equations. The three unknowns are the initial number (N_0), the growth rate parameter (r), and the carrying capacity (K). The appropriate general equation applied to each of the samples is

$$N_i = \frac{N_0 K_i^{rt_i}}{(K - N_0)(1 - e^{rt_i})} \quad (i = 1, 2, 3) \qquad (6\text{-}13)$$

where N_i is the number in the population at t_i. These equations yield the solution for carrying capacity, K, as

$$K = A/B \qquad (6\text{-}14)$$

where

$$A = (1/N_1) + (1/N_3) - (2/N_2)$$
$$B = (1/N_1 N_3) - (1/N_2)^2$$

and growth rate, r, is given as

$$r = \ln(C/D) \quad (6\text{-}15)$$

where

$$C = (1/N_2) - (1/N_1)$$
$$D = (1/N_3) - (1/N_2)$$

If we use population data for U.S. in 1870 = 40 million, in 1920 = 106 million, and in 1970 = 203 million, the ultimate population, according to this estimation method will be 324 million. This is in constrast to the Pearl and Reed (1920) estimate, which established the ceiling at 197 million (Keyfitz, 1985). There are problems with this approach, including changes in technology that continually alter the ceiling. Similar sorts of problems are encountered when this approach is applied to ecological problems.

Mortality Curve: The Gompertz

The Gompertz curve is frequently fitted to survival data expressed in one of two ways (Keyfitz, 1985)—as survival to age x (l_x) or as period mortality (q_x). The Gompertz equations for each case are

$$l_x = \exp\left[\frac{A_0}{G}(1 - e^{Gt})\right] \quad (6\text{-}16a)$$

$$q_x = A_0 e^{-Gx} \quad (6\text{-}16b)$$

The model requires two parameters: i) the initial mortality rate, A_0; and ii) the rate of sensescence, G. Their values can be determined by noting that the equation for q_x can be expressed as a linear equation by taking logs of both sides, yielding

$$\log_e q_x = -Gx + \log_e A_0 \quad (6\text{-}17)$$

Therefore a simple least squares fit of ln q_x vs. age x will yield values for the two parameters. As an example, logs of the non-zero q_x's for the medfly survival were plotted against age. The least squares fit resulted in the equation

$$y = .0581x - 6.9591$$

where y is ln(q_x) and x is age. Thus the parameters A_0 and G are

$$A_0 = \exp(-6.9591)$$
$$= .00095$$

and

$$G = .0581$$

As an example, l_x and q_x at age 45 are estimated by the Gompertz curve as

$$l_{45} = \exp[(.00095/.0581)(1 - e^{.0581 \times 45})]$$
$$= .81$$
$$q_{45} = q_0 \exp(-G \times 45)$$
$$= .00007$$

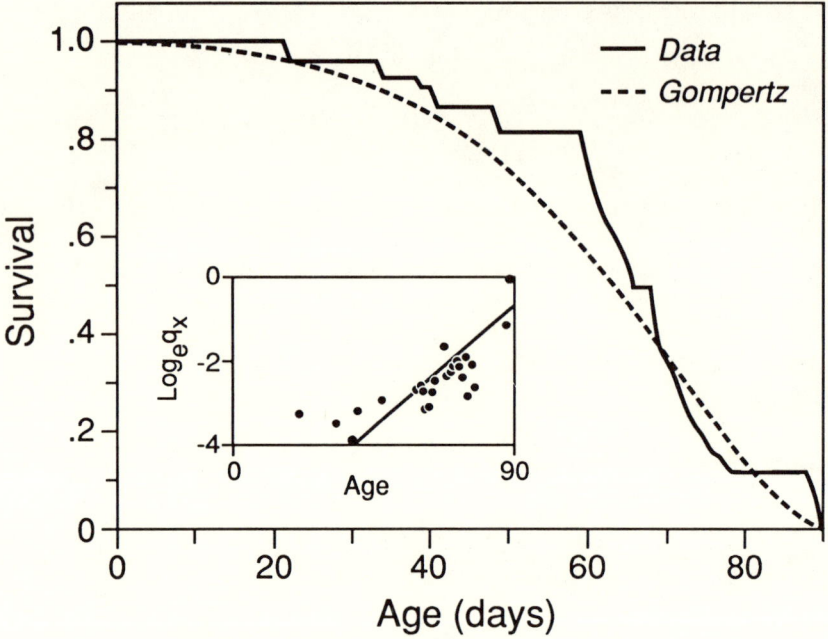

Figure 6-4. Gompertz curve fitted to medfly data.

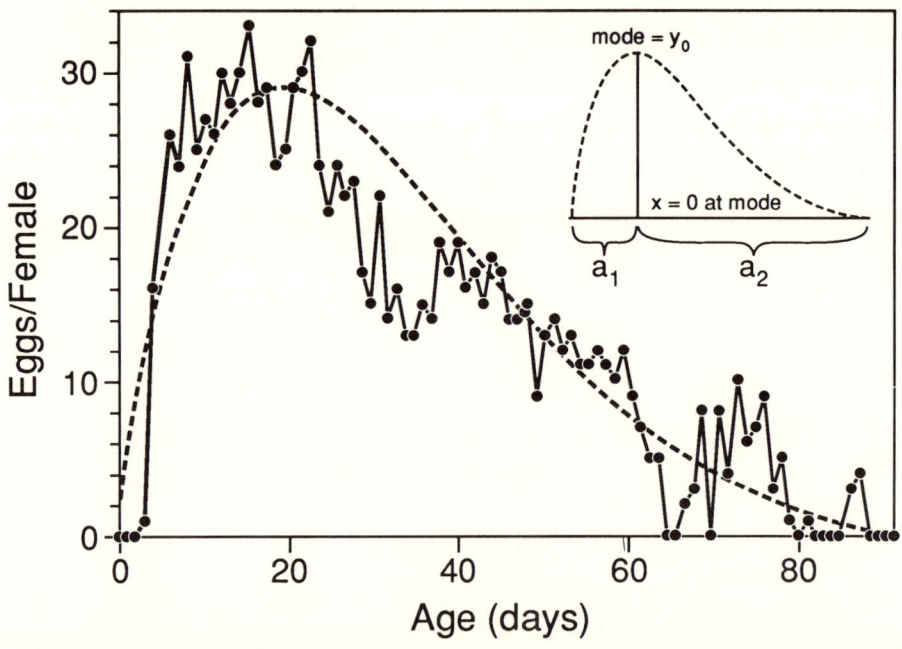

Figure 6-5. Pearson type I curve fitted to medfly fecundity data.

DEMOGRAPHIC APPLICATIONS

In general, two parameters are needed—G and A_0—to specify the Gompertz equation completely. The graph associated with this equation is given in Figure 6-4. Additional perspectives on the Gompertz equation are contained in Appendix 1.

Fecundity Curve: The Pearson Type I

Age-specific fecundity curves require three general properties in order to be useful in demography: i) potential to assume an asymetrical pattern: ii) ease with which fecundity level is adjusted; iii) insertion of preovipositional period or zero fecundity days. The Pearson type I curve with origin at the mode is of the form

$$y = y_0 \left(1 + \frac{x}{a_1}\right)^{m_1} \left(1 - \frac{x}{a_2}\right)^{m_2} \qquad (6\text{-}18)$$

A diagrammatic explanation of these parameters as well as an example fit is given in Figure 6-5. The fitting procedure consisted of approximating the parameters a_1 and a_2 by visual inspection of the fecundity curve to which the equation is to be fit and then using trial-and-error to obtain the best visual fit. The curve was fitted to the fecundity schedule of the medfly from Carey (1984) using $a_1 = 20$, $a_2 = 140$, $m_1 = 1$, $m_2 = 7$. Therefore $a_1/m_1 = a_2/m_2$. A closer fit can be obtained using numerical methods.

MASS REARING: BASIC HARVESTING CONCEPTS

The use of mass-reared insects in insect pest management dates back to the 1920s when parasites were cultured on a large scale and released for establishing or augmenting biological control programs. However, the chief impetus for developing large-scale rearing programs came as a result of the success of the sterile insect release method used against the screwworm, *Cochliomyia hominovorax*, in the southeastern U.S. The program required the production of 50 million flies per week for a period of 18 months. Subsequently this technique has been applied to many other insect species, including fruit flies, mosquitoes, the boll weevil, and several lepidopterans. The development of methods for large-scale rearing preceded implementation of each of these programs. Successful insect mass rearing involves solving biological problems in breaking diapause, developing diets, and establishing optimal temperature and humidity conditions. All programs must also be concerned with numerical considerations, including design of buildings for convenient monitoring of production and evaluation of insect quality, optimal adult discard age, harvest rates of the target stage, and size and age distributions of individuals released into the wild.

Harvesting theory was originally developed for use by wildlife and fisheries managers to help them determine which age classes to harvest to maintain the highest sustainable yields (Goodman, 1980). From a demographic perspective harvesting is concerned with cohort-structured populations (Getz and Haight, 1989) where the objective is to remove precisely the right number of individuals from the population to confer zero population growth. This is done simply by imposing a new survival schedule on the population via the harvesting (Beddington and Taylor, 1973).

Medfly

The general harvesting strategy applied to medflies is applicable to any species whose stages can be easily separated in rearing. This includes primarily the holometabolous species, which have the four stages of eggs, larvae, pupae, and adults. The Mediterranean fruit fly is often reared in massive numbers and will serve as the example for the development and illustration of this harvesting strategy (after Carey and Vargas, 1985).

Let θ denote the target age for harvesting and h denote the fraction of the individuals at the target age that are harvested. This leaves the fraction $(1 - h)$ for colony maintenance. The rate of harvest must confer zero population growth so the value of h must be the solution to the equation

$$1 = (1 - h) \sum_{x=\theta}^{\delta} L_x m_x \qquad (6\text{-}19a)$$

$$h = 1 - \left[\sum_{x=\theta}^{\delta} L_x m_x \right]^{-1} \qquad (6\text{-}19b)$$

where δ is the discard age of reproductives. Thus δ is now the artificially imposed last day of reproduction (and life). The complete demographic harvesting scheme is given in Figure 6-6.

The associated stable age distribution for the population within the rearing facility is

$$c_x = \begin{matrix} L_x/s & \text{for } x < \theta \\ L_x(1-h)/s & \text{for } x \geq \theta \end{matrix} \qquad (6\text{-}20)$$

where

$$s = \sum_{x=0}^{\theta-1} L_x + (1-h) \sum_{x=\theta}^{\delta} L_x$$

If c_θ is the proportion of the target stage at age θ in the colony after harvest, then the expression

$$P = \frac{2h}{1-h} \times \frac{c_\theta}{\sum_{x=\varepsilon}^{\delta} c_x} \qquad (6\text{-}21)$$

will give the daily per female yield, P, of the target stage, where ε denotes the age of eclosion. The 2 in the equation is needed to account for the production of males, assuming a 1:1 sex ratio. The derivation of this

DEMOGRAPHIC APPLICATIONS

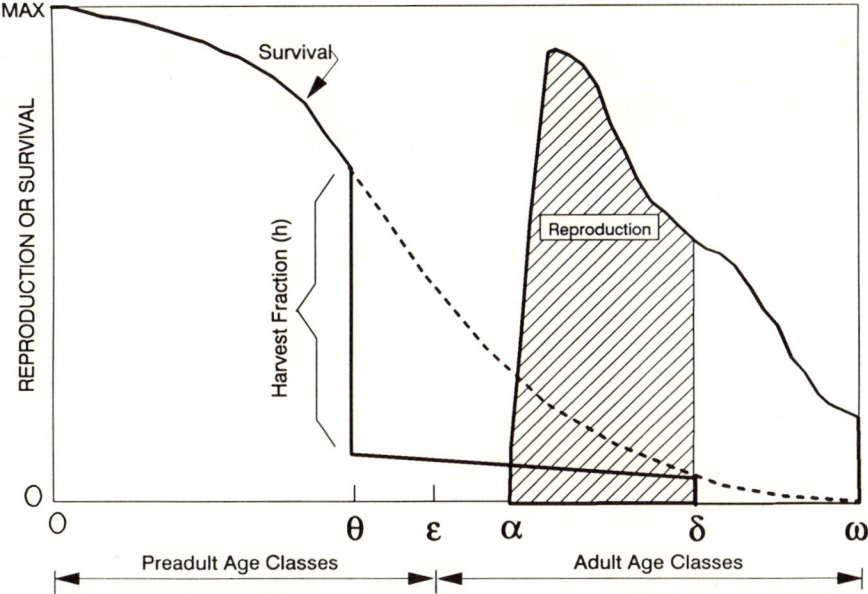

Figure 6-6. Schematic framework of the medfly mass-rearing problem where θ is harvest age, ε is age at eclosion, α is age of first reproduction, δ is discard age, ω is last possible age, and h is the fraction harvested (after Carey et al., 1985).

expression comes from the fact that the ratio of h to $1-h$ represents the ratio of harvested pupae to unharvested (remaining) pupae and the ratio of c_θ to $\sum c_x$ (summed from ε to δ) represents the ratio of unharvested pupae in the factory to the standing crop of adults. Therefore the product of these two ratios will yield the per adult production of the target stage.

Since all terms in this expression were originally given in terms of L_x and m_x, it can be shown that

$$\frac{h}{1-h} = \left[\sum_{x=0}^{\delta} L_x m_x \right] - 1 \qquad (6\text{-}22)$$

and

$$\frac{c_\theta}{\sum_{x=\varepsilon}^{\delta} c_x} = \frac{L_\theta}{\sum_{x=\varepsilon}^{\delta} L_x} \qquad (6\text{-}23)$$

Therefore the expression for P can be given in terms of the original vital rates without having to compute h:

$$P = \frac{2L_\theta \left[\sum_{x=0}^{\delta} (L_x m_x) - 1 \right]}{\sum_{x=\varepsilon}^{\delta} L_x} \qquad (6\text{-}24)$$

This is a particularly useful form since per-female productivity of the target stage can be computed directly from the survival schedules and the cumulative net maternity schedule (summed to the prospective discard age, δ) of the unharvested population.

As an example application consider the survival and reproductive schedules for the medfly given in Table 6-12.

The complete harvesting evaluation requires three steps:

Step 1. Harvest rate. The fraction, h, that must be harvested for a discard age of 30 is $h = 1 - (1/62.3) = .9839$; for age 35 it is $h = 1 - (1/91.25) = .9890$, and for age 40 it is $h = 1 - (119.76) = .9917$.

Step 2. Factory age structure. Using a target age, θ, of 20 and a discard age, δ, of 40 days the denominator s is computed as

$$s = \sum_{x=0}^{18} L_x + (1-h) \sum_{x=19}^{40} L_x$$

$$= 12.34 + (1 - .9917)11.45$$

$$= 12.43$$

Therefore the fraction in the egg stage is $(.99 + .98)/s = .158$. The fraction in the larval, pupal, and adult stages can be determined by summing the L_x values over the appropriate age classes and dividing by s.

Step 3. Production rate. Substituting $L_\theta = .55$ (survival to target stage age $x = 19$), 119.76 for net reproduction to discard age, and 10.89 for the denominator of Equation 6-24 yields a per female production rate, P, of 12.00 pupae/female/day.

Spider Mites

Tetranychid mites are typically mass-reared by infesting pots of flats of young host plants such as beans. After a week or two, depending on the initial infestation rate, a fraction of mites from a fraction of all flats are removed (harvested) and the remainder used for reinfesting new plants. The harvesting problem for spider mites is simple conceptually in that a fraction of the whole population is harvested and the remainder is left for renewal. The objective is to harvest the appropriate number to confer population replacement.

The appropriate harvesting model for this case is (see Carey and Krainacker, 1988)

$$1 = \sum_{x=0}^{\omega} (1-h)^x \lambda^{-x} L_x m_x \qquad (6\text{-}25)$$

$$h = (\lambda - 1)/\lambda$$

$$= 1 - \lambda^{-1}$$

where $\lambda = e^r$. Here h is the fraction of each age class that is removed daily. Note that the harvest rate for the model is 1 minus the inverse of the per

Table 6-12. Demographic Parameters for Medfly Harvesting Model

Age x	Survival L_x	Reproduction m_x	Net Reproduction $L_x m_x$	Cumulative Net Reproduction $L_x m_x$	Cumulative Survival L_x	Harvest Fraction h	Production Rate P
Egg							
0	.99					—	—
1	.98					—	—
Larva							
2	.90					—	—
3	.79					—	—
4	.72					—	—
5	.65					—	—
6	.61					—	—
7	.59					—	—
8	.58					—	—
Pupa							
9	.57	.00	.00	.00	—	—	—
10	.56	.00	.00	.00	—	—	—
11	.55	.00	.00	.00	—	—	—
12	.55	.00	.00	.00	—	—	—
13	.55	.00	.00	.00	—	—	—
14	.55	.00	.00	.00	—	—	—
15	.55	.00	.00	.00	—	—	—
16	.55	.00	.00	.00	—	—	—
17	.55	.00	.00	.00	—	—	—
18	.55	.00	.00	.00	—	—	—
19	.55	.00	.00	.00	—	—	—
Adult							
20	.54	.00	.00	.00	.54	—	—
21	.54	.00	.00	.00	1.08	—	—

Table 6-12. (Contd.)

Age x	Survival L_x	Reproduction m_x	Net Reproduction $L_x m_x$	Cumulative Net Reproduction $L_x m_x$	Cumulative Survival L_x	Harvest Fraction h	Production Rate P
22	.53	.00	.00	.00	1.61	—	—
23	.52	9.19	4.78	4.78	2.13	.7907	1.95
24	.52	16.56	8.61	13.39	2.65	.9253	5.14
25	.52	17.07	8.88	22.27	3.17	.9551	7.38
26	.52	17.38	9.04	31.30	3.69	.9681	9.03
27	.52	18.34	9.54	40.84	4.21	.9755	10.41
28	.52	14.10	7.33	48.17	4.73	.9792	10.97
29	.52	15.98	8.31	56.48	5.25	.9823	11.62
30	.52	11.18	5.81	62.30	5.77	.9839	11.69
31	.52	12.92	6.72	69.01	6.29	.9855	11.89
32	.52	12.30	6.40	75.41	6.81	.9867	12.02
33	.52	11.15	5.80	81.21	7.33	.9877	12.04
34	.52	11.18	5.81	87.02	7.85	.9885	12.05
35	.52	8.14	4.23	91.25	8.37	.9890	11.86
36	.52	13.87	7.14	98.40	8.88	.9898	12.06
37	.51	11.92	6.08	104.48	9.39	.9904	12.12
38	.51	11.00	5.56	110.03	9.90	.9909	12.11
39	.50	8.92	4.46	114.49	10.40	.9913	12.00
40	.49	10.76	5.27	119.76	10.89	.9917	12.00
41	.48	7.47	3.55	123.31	11.37	.9919	11.84
42	.47	9.31	4.38	127.69	11.84	.9922	11.78
43	.47	5.34	2.51	130.20	12.31	.9923	11.55
44	.46	5.87	2.67	132.87	12.76	.9925	11.37
45	.44	9.56	4.21	137.08	13.20	.9927	11.34
46	.43	10.15	4.31	141.39	13.63	.9929	11.33
47	.41	8.55	3.46	144.85	14.03	.9931	11.28
48	.39	8.88	3.42	148.27	14.42	.9933	11.24

49	.36	6.14	2.18	150.45	14.77	.9934	11.13
50	.34	6.57	2.23	152.68	15.11	.9935	11.04
51	.31	5.77	1.76	154.44	15.42	.9935	10.95
52	.24	7.03	1.69	156.13	15.66	.9936	10.90
53	.19	4.84	.92	157.05	15.85	.9936	10.83
54	.17	6.67	1.13	158.18	16.02	.9937	10.80
55	.17	5.34	.88	159.07	16.18	.9937	10.75
56	.15	6.32	.95	160.01	16.33	.9938	10.71
57	.14	3.90	.53	160.54	16.47	.9938	10.66
58	.12	3.00	.36	160.90	16.59	.9938	10.61
59	.11	3.63	.38	161.28	16.69	.9938	10.56
60	.08	2.93	.23	161.52	16.77	.9938	10.53
61	.06	5.25	.29	161.80	16.83	.9938	10.51
62	.05	6.00	.30	162.10	16.88	.9938	10.50
63	.05	2.25	.10	162.21	16.92	.9938	10.48
64	.04	2.75	.11	162.32	16.96	.9938	10.46
65	.04	9.75	.39	162.71	17.00	.9939	10.46
66	.04	4.00	.16	162.87	17.04	.9939	10.45
67	.04	9.00	.32	163.18	17.08	.9939	10.45
68	.03	3.50	.11	163.29	17.11	.9939	10.44
69	.03	3.25	.10	163.38	17.14	.9939	10.42
70	.03	1.25	.04	163.42	17.17	.9939	10.41
71	.03	5.50	.14	163.56	17.19	.9939	10.40
72	.02	3.00	.06	163.62	17.21	.9939	10.39
73	.02	2.50	.05	163.67	17.23	.9939	10.39
74	.02	7.00	.11	163.77	17.25	.9939	10.38
75	.01	4.50	.02	163.80	17.25	.9939	10.38

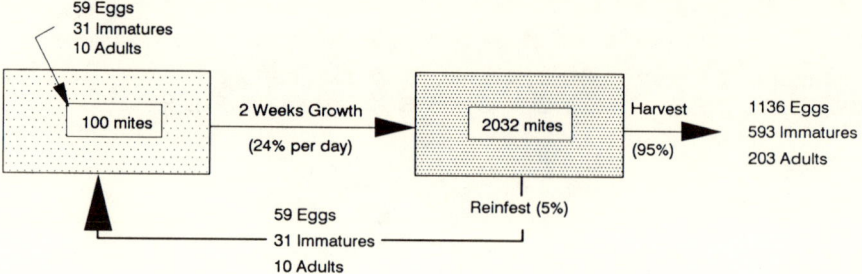

Figure 6-7. Diagrammatic sketch of 2-week mass-production cycle for the two-spotted spider mite (after Carey and Krainacker, 1988).

generation rate of change (net reproductive rate), while harvest rate in the second model is 1 minus the inverse of the daily rate of increase (λ). A simple extension for the second model regards the fraction removed after t days of population growth, denoted h_t:

$$h_t = 1 - \lambda^{-t} \tag{6-26}$$

Because the same fraction of each age class is removed from the population, the age structure of the factory population will be identical with the stable age distribution for the unconstrained (unharvested) case. The per capita production rate will simply be h or h_t. If we use a rate of increase of $\lambda = 1.24$ for *T. urticae* the harvest rate, h, is

$$h = 1 - 1.24^{-1}$$
$$= .194$$

and the two-week harvest rate is

$$h_{14} = 1 - 1.24^{-14}$$
$$= .951$$

These results state that 19.4% of all stages of mites can be harvested daily or, more practically, 95.1% of all mites can be harvested from flats where mite populations are allowed to grow for 2 weeks. Thus out of a hundred 2-week-old mite populations on infested flats, one of two harvesting strategies can be employed: i) remove 5% of all mites from each flat for reinfesting the next 100 flats; or ii) use the mites from five full flats to reinfest the next hundred flats. A sketch of this strategy is given in Figure 6-7.

PARASITOID MASS REARING

Parasitoid rearing departs demographically from the classical harvesting case in several fundamental ways (Carey et al., 1988). *First*, the target sex for parasitoids is the female rather than the male, as is the case for the sterile

DEMOGRAPHIC APPLICATIONS

male technique. *Second*, most parasitoids are haplodiploid hymenopterans where sex ratios are seldom unity. Therefore the various determinants of sex ratios must be fully understood in the parasitoid species to be reared. *Third*, artificial diets have not been developed for the vast majority of parasitoids, thus hosts must be mass reared in order to rear parasitoids. Consequently the logistics of parasitoid production is much more involved than single-species rearing. *Fourth*, to optimize production of female parasitoids, the scale of host production must be such that the appropriate number of hosts of the right stage are available (harvested) for exposure to the standing crop of female parasitoids. Therefore efficient production requires an appropriate balance of hosts and parasitoids.

My specific objective in this section is to develop a formal demographic framework for addressing several questions related to insect parasitoid mass rearing, including i) at what age should parasitoid females used in the production stock be discarded to optimize production of harvestable females? ii) how many hosts are required to produce one harvestable female parasitoid? and iii) what is the age, sex, and species structure of a balanced system? I use *Biosteres tryoni*, a braconid parasitoid of the Mediterranean fruit fly, *Ceratitis capitata*, as a model, but the techniques are general.

Harvest Rates

Parasitoid males and females are treated separately in the model since their primary sex ratios vary with maternal age. Let L_x^f and L_x^m denote survival to age x of parasitoid females and males, respectively, and m_x^f and m_x^m denote a female parasitoid's production of female and male offspring at age x, respectively. Therefore the per female production of daughters, denoted R_o^f, and sons, denoted R_o^m, from their age of first reproduction to their discard age (δ) is given as

$$R_o^f = \sum_{x=\alpha}^{\delta} L_x^f m_x^f \qquad (6\text{-}27a)$$

$$R_o^m = \sum_{x=\alpha}^{\delta} L_x^f m_x^m \qquad (6\text{-}27b)$$

Hence the harvest rate, h_f, of female parasitoids required to confer zero population growth in their population must be the solution to the equation

$$1 = (1 - h_f) R_o^f \qquad (6\text{-}28a)$$

$$h_f = 1 - (R_o^f)^{-1} \qquad (6\text{-}28b)$$

and the primary sex ratio (male:female), denoted S, is

$$S = R_o^m / R_o^f \qquad (6\text{-}29)$$

The harvest rates necessary to confer zero population growth for both the medfly and *Biosteres* for different age classes are given in Table 6-5. Perhaps the most revealing aspect of these relations is the enormous difference between

Table 6-13. Net Reproduction (Female Offspring) by Age for the Mediterranean Fruit Fly and the Parasitoid, B. tryoni, and the Associated Harvest Rates, h_h and h_f

Age Class	C. capitata		B. tryoni	
	Net Reproduction	Harvest Rate (h_h)	Net Reproduction	Harvest Rate (h_f)
20	.00	—	.00	—
25	3.61	.723	13.40	.925
30	51.94	.981	19.89	.950
35	86.32	.988	24.12	.959
40	104.16	.990	24.84	.960

the changes in net reproduction rates with age between the species. From age class 25 to 40 net reproduction for *Biosteres* increases by less than 2-fold from 13.4 to 24.8, while net reproduction for the medfly increases by almost 30-fold, from 3.6 to 104.2 (Table 6-13). The specific harvest rates do not reflect the optimal discard ages since the objective in mass rearing is not to maximize the fraction harvested but rather to maximize the per capita production rates for adult females (see next section). Harvest rates, h_h and h_f, simply give the numerical requirements for exact replacement in the medfly host and the female parasitoid, respectively.

Per Female Production Rates

The per female production rate of the host target stage, denoted P_h, is

$$P_h = \frac{2(R_o - 1)L_\theta}{\sum_{x=\varepsilon}^{\delta} L_x} \quad (6\text{-}30)$$

where θ denotes the age at which the target stage is harvested and ε represents eclosion age. The interpretation of Equation 6-30 is this: The term $(R_o - 1)$ gives the per female number of hosts to be removed (harvested) for confering exact replacement. Therefore this term multiplied by L_θ $[2(R_o - 1)L_\theta]$ yields the per female number of female hosts age θ to be removed for exact replacement. This product divided by the number of female-days in the standing cohort aged ε to δ (i.e., the denominator in eq. 6-30) yields the number of females aged θ to be removed for exact replacement. The "2" in the numerator corrects for male offspring that may be harvested.

Following the same reasoning, the per capita adult female parasitoid rates for production of harvestable daughters and sons, denoted P_f and P_m, respectively, are

$$P_f = \frac{[R_o^f - 1]L_\theta^f}{\sum_{x=\varepsilon}^{\delta} L_x^f} \quad (6\text{-}31a)$$

$$P_m = P_f S \quad (6\text{-}31b)$$

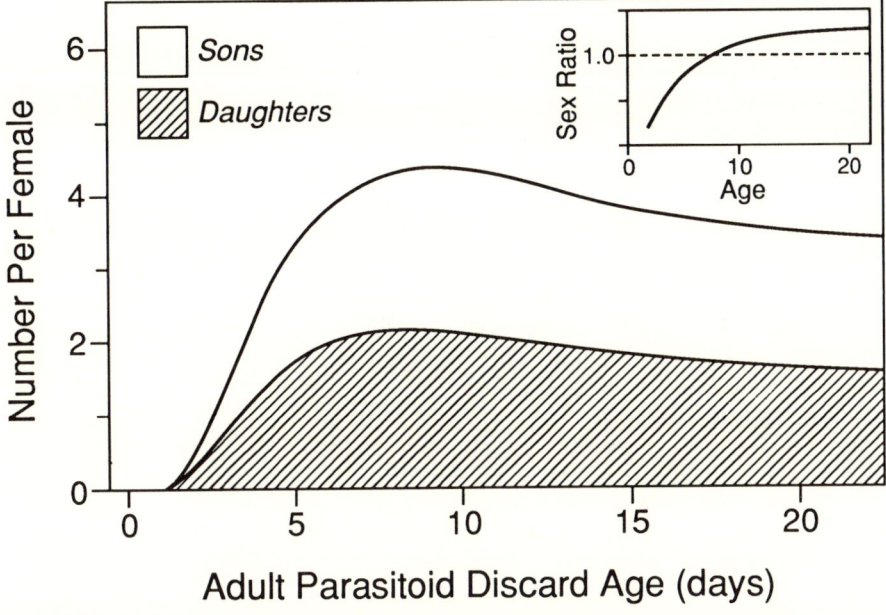

Figure 6-8. Per capita adult female *B. tryoni* production rates of male and female offspring reared on medfly hosts (after Carey et al., 1988).

Production rates of medfly larvae per adult medfly female ranged from about 7 at early medfly discard ages to over 18 at older medfly discard ages. These rates were similar to those for per capita production rates of medfly pupae given in Carey and Vargas (1985). Thus the main focus here will be on *Biosteres* production.

The trade-offs between discard age (δ) and per capita female parasitoid production rates of harvestable male (P_m) and female (P_f) offspring are given in Figure 6-8, including the male to female ratio, S (Figure 6-8, inset). Three aspects of this figure merit comment. *First*, the maximum total production rate over all discard ages is only slightly over 4 (both sexes) per female/day, only two of which are daughters. This is considerably lower than the per capita rate for the medfly target stage, P_h, which was about 15 (both host sexes) per female/day. These differences are due entirely to the differences in net reproduction between the species. *Second*, the optimal adult *Biosteres* per capita production is relatively insensitive to discard age as reflected in the broad peak ranging from ages 6 to 14 days. As long as the discard age is close to the theoretical optimum, the rearing facility will produce harvestable parasitoids at nearly the maximum rate (e.g., Plant, 1986). This allows flexibility in setting operational parameters. *Third*, the primary sex ratio (S) varies from about .6 (63% females) at low δ to about 1.15 (47% females) at high δ. Thus lower δ's are needed if the rearing objective is to maximize the percentage of harvestable females rather than simply the number of parasitoid females. On the other hand, very large δ's reduce both per capita number of harvestable females and the percentage females.

King (1987) lists 14 factors that may affect offspring sex ratios in parasitoid wasps, most of which are relevant to the current system. However, we do not explicitly address each of these factors or incorporate them in our analysis. Instead we assume that the empirically derived sex ratios used to parameterize the system will remain constant as long as the rearing conditions are kept the same. For example, host and parasitoid female densities both affect offspring sex ratios. We assume that rearing densities in the factory will be identical to those under which the sex ratios were derived. Curve fitting to incorporate a density effect in the model would add little conceptually, though it might be of practical value.

Host-Parasitoid Coupling

Conventional life tables typically consider survival to age x as a fraction of the number of newborn. The normalizing term associated with this fraction is the life table radix and is usually unity. In a coupled system such as ours the radices of both species must be re-scaled. This includes separate radices for the female and male parasitoids since female offspring represent a different fraction of total newborn than do male offspring.

There are two reasons why this re-scaling is necessary. *First*, the number of hosts needed for parasitization is not relative to host females (or males) but rather to parasitoid females. *Second*, the overall age, species, and sex composition of a mass-rearing factory cannot be determined by considering each species or sex in isolation.

The normalization of the radices for both the host and the parasitoid by sex is determined as follows. The number of hosts needed per parasitoid female is

$$\frac{R_o^f + R_o^m}{\sum_{x=\varepsilon}^{\delta} L_x^f} \qquad (6\text{-}32)$$

The numerator in Equation (6-32) gives the number of hosts required by the average female in her lifetime, and the denominator gives the average number of days lived by the average female.

The expression $h_h L_\theta$ gives the fraction of newborn hosts that are harvested when they attain age θ. It also represents the number of the target stage harvested *per newborn individual*. Therefore Equation 6-32 (i.e., number in target stage harvested *per female parasitoid*) divided by this expression yields the *number of newborn hosts per female parasitoid*:

$$\frac{R_o^f + R_o^m}{h_h L_\theta \sum_{x=\varepsilon}^{\delta} L_x^f} \qquad (6\text{-}33)$$

DEMOGRAPHIC APPLICATIONS

The number of newborn female parasitoids needed per adult female parasitoid in the standing crop to insure replacement subject to their harvest rate h_f is given by

$$\frac{1}{(1-h_f)\sum_{x=\varepsilon}^{\delta} L_x^f} \tag{6-34}$$

This term gives the *number of newborn parasitoid females per adult parasitoid female*.

Since Equation 6-33 gives the newborn hosts/adult parasitoid female and Equation 6-34 gives the newborn female parasitoids/adult parasitoid female, the dividend of these two equations (i.e., Eq. 6-33/Eq. 6-34) yields the number of newborn host (both sexes) per newborn female parasitoid. This is the non-normalized radix of the host life table *relative* to the non-normalized radix of the parasitiod female table. If we use D to denote this ratio, the dividend reduces to

$$D = \frac{(1-h_f)(R_o^f + R_o^m)}{h_h L_\theta} \tag{6-35}$$

This expression gives the radix value for the life table of *both* host sexes. That is, the radix for either sex alone equals half of D (i.e., D/2). Since parasitoid female newborn are used as reference, the radix for the female parasitoid newborn is unity ($=1$) and the radix for the male parasotoid newborn equals S.

The basic parameters L_x and m_x for the host and L_x^f and m_x^f for the parasitoid before harvest are given in Table 6-14 by age and stage in the columns headed "unharvested". The number of adult parasitoid female-days is computed as

$$\sum_{x=19}^{34} L_x^f = .68 + .66 + .64 + \cdots + .48$$

$$= 7.49$$

The number of hosts needed per parasitoid female is (from Eq. 6-32)

$$\frac{R_o^f + R_o^m}{7.49} = \frac{23.24 + 25.06}{7.49}$$

$$= 6.45 \text{ mid-3rd instar hosts/adult } Biosteres \text{ female}$$

This number divided by the number of harvested hosts per newborn host ($=h_h L_\theta$) yields the number of newborn hosts per female parasitoid:

$$\frac{6.45}{h_h L_\theta} = \frac{6.45}{(.988)(.71)}$$

$$= 9.19 \text{ newborn hosts/adult } Biosteres \text{ female}$$

Table 6-4. Survival and Reproductive Schedules of *C. capitata* and *B. tryoni* in the Absence of Harvest and with Harvest for Both Species in the Coupled Rearing System[1]

Age Class	*C. capitata*						*B. tryoni*					
	Unharvested			Harvest (h_h = .988)				Unharvested			Harvest (h_f = .957)	
	Stage	L_x	m_x	L_x	$L_x m_x$		Stage	L_x^f	m_x^f		L_x^f	$L_x^f m_x^f$
0	Egg	1.00	.00	1.4804	.0000		Egg	1.00	.00		1.0000	.0000
1	Egg	.90	.00	1.3323	.0000		Egg	.98	.00		.9800	.0000
2	Larva	.85	.00	1.2583	.0000		Larva	.96	.00		.9600	.0000
3	Larva	.80	.00	1.1843	.0000		Larva	.94	.00		.9400	.0000
4	Larva	.75	.00	1.1103	.0000		Larva	.92	.00		.9200	.0000
5	Larva	.74	.00	1.0955	.0000		Larva	.90	.00		.9000	.0000
6	Larva	.73	.00	1.0807	.0000		Larva	.88	.00		.8800	.0000
7	Larva	.72	.00	1.0659	.0000		Larva	.86	.00		.8600	.0000
8	(*I*)	.71	.00	.0126	.0000		Larva	.85	.00		.8500	.0000
9	Pupa	.70	.00	.0124	.0000		Pupa	.83	.00		.8300	.0000
10	Pupa	.70	.00	.0124	.0000		Pupa	.81	.00		.8100	.0000
11	Pupa	.69	.00	.0123	.0000		Pupa	.79	.00		.7900	.0000
12	Pupa	.69	.00	.0123	.0000		Pupa	.78	.00		.7800	.0000
13	Pupa	.68	.00	.0121	.0000		Pupa	.76	.00		.7600	.0000
14	Pupa	.68	.00	.0121	.0000		Pupa	.75	.00		.7500	.0000
15	Pupa	.68	.00	.0121	.0000		Pupa	.73	.00		.7300	.0000
16	Pupa	.67	.00	.0119	.0000		Pupa	.71	.00		.7100	.0000
17	Pupa	.66	.00	.0117	.0000		Pupa	.70	.00		.7000	.0000
18	Pupa	.66	.00	.0117	.0000		Pupa	.70	.00		.7000	.0000
19	Pupa	.65	.00	.0115	.0000		Adult(h)	.68	.00		.0292	.0000
20	Pupa	.65	.00	.0115	.0000		Adult(h)	.66	.00		.0284	.0000
21	Adult	.65	.00	.0115	.0000		Adult(h)	.64	1.60		.0275	.0440

22	Adult	.65	.00	.0115	.0000	Adult(h)	.62	3.60	.0267	.0960
23	Adult	.63	.00	.0112	.0000	Adult(h)	.60	5.30	.0258	.1367
24	Adult	.62	.00	.0110	.0000	Adult(h)	.59	6.40	.0254	.1624
25	Adult	.60	6.01	.0109	.0656	Adult(h)	.57	5.60	.0245	.1373
26	Adult	.58	14.70	.0106	.1551	Adult(h)	.56	4.70	.0241	.1132
27	Adult	.57	17.81	.0104	.1847	Adult(h)	.55	4.10	.0237	.0970
28	Adult	.55	17.85	.0100	.1786	Adult(h)	.53	3.40	.0228	.0775
29	Adult	.52	19.80	.0095	.1873	Adult(h)	.51	2.70	.0219	.0592
30	Adult	.50	19.17	.0091	.1744	Adult(h)	.50	2.10	.0215	.0452
31	Adult	.49	17.13	.0089	.1527	Adult(h)	.48	1.50	.0206	.0310
32	Adult	.48	16.07	.0087	.1403	(δ)	23.34	41.00	16.1721	1.0000
33	Adult	.47	15.14	.0085	.1294					
34	(δ)	.45	13.68	.0082	.1120					
		23.07	157.36	9.9043	1.4804					

The number of newborn parasitoid females per adult parasitoid female is

$$\left[(1-h_f)\sum_{x=19}^{34} L_x^f\right]^{-1} = [(1-.957)7.49]^{-1}$$

$$= 3.105 \text{ newborn } \textit{Biosteres} \text{ females/adult } \textit{Biosteres} \text{ female}$$

Thus the number of newborn hosts (either sex) per newborn $\textit{Biosteres}$ female is

$$9.193/2(3.105)$$

$$= 1.4804 \text{ newborn medfly/newborn } \textit{Biosteres} \text{ female}$$

The 2 in the denominator is needed to account for both host sexes. The overall interpretation is that for every newborn $\textit{Biosteres}$ female in the factory standing population there will be 1.4804 newborn host females, the same number of host males, and 1.079 newborn male parasitoids. These represent the appropriate life table radices for both the host and the parasitoid in the mass-rearing factory.

The survival column for the harvested host population therefore begins with the radix = 1.4804 (see Table 6-14) of which the fraction .72, or 1.0659 individuals, survive to age 7 days. The fraction .71 of the original radix survive to age 8 (= age θ) but of these the fraction h = .988 are harvested, leaving .0126 for host renewal. Note that the net replacement rate for the host (sum of harvested $L_x m_x$ column) is 1.4804, which is the same as the host radix.

The same methods are applied to the $\textit{Biosteres}$ factory life table using the harvest rate $h_f = .957$ at age 19 days. In this case the net replacement rate (sum of $L_x^f m_x^f$ column) is unity since the radix for $\textit{Biosteres}$ females is used for normalizing the overall system.

Age and Species Structure

Expressions for the fraction of the total factory population (host + parasitoid) aged x for female hosts, denoted c_x, and for the female parasitoids, denoted c_x^f, are

$$c_x = \begin{cases} \dfrac{DL_x}{V} & x < \theta \\ \dfrac{D(1-h_h)L_x}{V} & x \geq \theta \end{cases} \quad (6\text{-}36a)$$

$$c_x^f = \begin{cases} \dfrac{L_x^f}{V} & x < \theta \\ \dfrac{(1-h_f)L_x^f}{V} & x \geq \theta \end{cases} \quad c_x^f \quad (6\text{-}36b)$$

DEMOGRAPHIC APPLICATIONS

where

$$V = 2D \sum_{x=0}^{\theta-1} L_x + [1+(1/S)] \sum_{x=0}^{\theta-1} L_x^f + 2D(1-h_h) \sum_{x=\theta}^{\delta} L_x$$

$$+ (1-h_f)[1+(1/S)] \sum_{x=\theta}^{\delta} L_x^f \qquad (6\text{-}36c)$$

The fraction of the host male population aged x equals that of the female and the fraction of the total factory population aged x in the male parasitoid population is

$$c_x^m = Sc_x^f \qquad (6\text{-}37)$$

I assume that medfly larvae become part of the *Biosteres* population immediately on becoming parasitized. However, parasitized medfly larvae do not typically die of parasitization until the pupal stage.

A specific numerical example of the host-parasitoid mass-rearing system is used to illustrate techniques for coupling the system and, in turn, for computing overall factory stage structure. I use a *Biosteres* harvest age of $\theta = 19$ days (newly eclosed adults); survival to age θ as $L_\theta = .68$; and a discard age of $\delta = 31$ days ($= 13$ days of adult life).

The net reproductive rate for the medfly host to $\delta = 34$ was $R_o = 81.38$, which gives a host target stage harvest rate of $h_h = .988$ [$= 1 - (81.38)^{-1}$]. The net reproductive rate for *Biosteres* females to age δ was $R_o^f = 23.24$, which gives a *Biossteres* harvest rate of $h_f = .957$. The net reproductive rate for male offspring was $R_o^m = 25.06$, resulting in a primary sex ratio of $S = 1.08$ ($= 25.06/23.24$) or one female offspring for every 1.079 male offspring.

Stage Structure

Once the overall system coupling is established through the normalization of radices, determining the stage structure of the rearing factory is simply a matter of finding the number of insect-days lived by both species and sexes in each of their cohorts. The sum total of all insect days lived by the host population is the sum of the L_x column (harvested) multiplied by 2 (to include males):

$$= 2(9.9043) = 19.809$$

and the number of insect-days for female *Biosteres* is the sum of their L_x column (harvested), which is 16.1721. The number of insect-days for male parasitoids is the S multiplied by this number:

$$= 1.079(16.1721) = 17.4497$$

Thus the sum total of all insect days in the factory equals S, which is

$$19.809 + 16.1721 + 17.4497 = 53.4308$$

Therefore the fraction of both the male and female host and parasitoid by

Table 6-15. Percentage of Each Stage in the Mass Rearing Factory for the Parasitoid, B. tryoni, and Its Medfly Host, C. capitata

Stage	C. capitata			B. tryoni		
	Female	Male	Total	Female	Male	Total
Egg	5.27	5.27	10.54	3.71	4.00	7.71
Larvae	12.77	12.77	25.54	11.83	12.77	24.60
Pupae	.27	.27	.54	14.17	15.30	29.47
Adults	.18	.18	.36	.60	.65	1.25
Total	18.49	18.49	36.98	30.31	32.72	63.03

stage in the factory is the sum of the insect-days spent in each stage divided by the total insect-days in the factory ($= 53.4308$). For example, the fraction of female host eggs is simply the sum of L_0 and L_1 for the harvested host life table:

$$\frac{1.4804 + 1.3323}{53.4308} = .0526$$

Thus female host eggs will represent 5.26% of all individuals in the factory. Since male host eggs equals this same percentage, a total of 10.52% of all individuals in the factory will consist of host eggs. Likewise, the fraction of all parasitoid female eggs is

$$\frac{1.0000 + .9800}{53.4308} = .0371$$

and that for parasitoid male eggs is

$$.0371 \, S = .0371(1.079) = .0400$$

The fraction of the overall factory population in any given stage for either the host or the parasitoid is conditional on the species; their vital rates; the stages selected for harvesting; and, in turn, the specific harvest rates and discard ages for both. Changes in any one of these elements will affect the overall structure of the factory populations; thus no standard structure exists. Rather than examine a wide range of parameter combinations, we focus on the specific numerical example given in the previous section.

Completion of the numerical computations for the medfly/parasitoid system given in the previous example yields the overall stage structure of the factory (Table 6-15) as well as the within-species partitioning (Figure 6-9). Several aspects of this particular structuring merit comment. *First*, the overall ratio of parasitoid to medflies is nearly 2 to 1. That is, 63% of the total factory population consists of *Biosteres* and 37% of medflies. This is despite the nearly 1.5:1 ratio of newborn medflies to newborn *Biosteres* as reflected in the normalized radices. This difference results largely from differences in the harvest ages between the two species. *Second*, the adult *Biosteres*

DEMOGRAPHIC APPLICATIONS

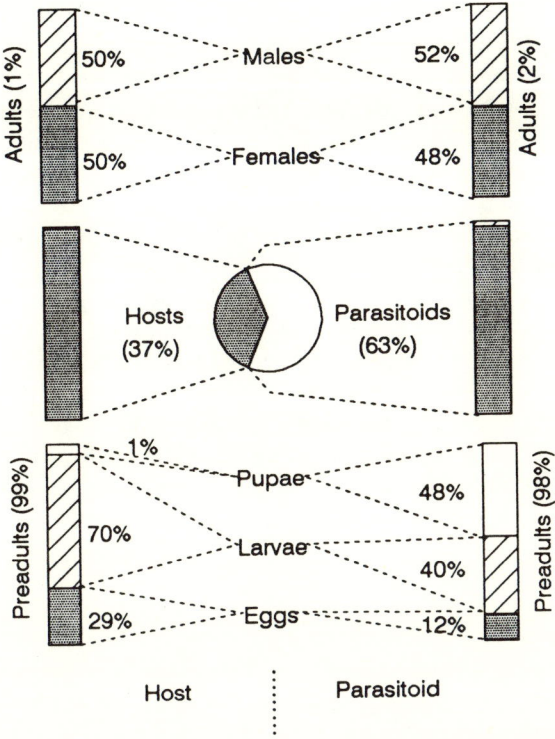

Figure 6-9. Stage structure of a parasitoid mass-rearing factory using target ages and harvest rates for *B. tryoni* and its medfly host given in Table 6-15 (after Carey et al., 1988).

population is 3.5-fold greater than the adult medfly population, but both constitute only about 1.6% of the overall factory population. *Third*, comparison of the within-species stage structures (i.e., percentage of own species population) reveals several striking differences that are due mostly to differences in stages harvested. For example, *Ceratitis* pupae constitute only about 1% of the preadult population, whereas pupae constitute 48% of the *Biosteres* population. The greatest similarities exist in the adult populations, which constitute less than 2% of the total for both species, and the sex ratios are equal to unity or close.

An additional aspect of these relations concerns the production rates or ratios for harvested female parasitoids relative to other stages in the factory rather than simply to the female adult parasitoid. For example, the fraction of the total *Biosteres* population that are adult females is .01 (= .0060/.6303). Since the production rate of harvestable female parasitoids per adult female parasitoid is roughly 2 (see Figure 6-8), there are approximately .02 harvested female parasitoids per individual parasitoid in the overall *Biosteres* population. In other words, 50 *Biosteres* exist in the factory standing population for every *Biosteres* female harvested. The same method reveals that there are 6.7 harvested parasitoid females per medfly adult female, 3.25 third instar medfly larvae per harvested parasitoid female, and 4.6 newborn medflies (i.e., new eggs) per harvested parasitoid female. The last rate can be viewed as the conversion ratio of newborn medflies to the ultimate target stage (i.e., parasitoid females) and represents an upper limit for this aspect of factory efficiency. That is, we assumed 100% parasitization. This parasitization level is seldom attained in large-scale production, and a more realistic conversion may be around seven to ten newborn medflies required for each parasitoid female that is ultimately harvested.

The assumption that parasitized medfly larvae and pupae are part of the *Biosteres* population greatly simplifies computation of age and species structure but distorts the overall numerical role of the medfly in parasitoid rearing. That is, all preadult *Biosteres* are contained wthin live medflies early in life and within medfly cadavers (puparia) later in life. Since the only parasitoids that are completely separate from the medfly are the adults and these constitutes only about 1-2% of the 2-species population, this means that rearing the other 98% of individuals involves rearing or holding medflies.

BIBLIOGRAPHY

Abbot, W. S. (1925) A method of computing the effectiveness of an insecticide. *J. Econ. Entomol.* 18:265–267.

Back, E. A., and C. E. Pemberton (1916) Effect of cold-storage temperatures upon the Mediterranean fruit fly. *J. Agric. Res.* 5:657–666.

Beddington, J. R., and D. B. Taylor (1973) Optimum age-specific harvesting of a population. *Biometrics* 29:801–809.

Bellows, T. S., and M. H. Birley (1981) Estimating developmental and mortality rates and stage recruitment from insect stage-frequency data. *Res. Popul. Ecol.* 23:232–244.

Braner, M., and N. G. Hairston, Jr. (1988). From cohort data to life table parameters via stochastic modeling. In: *Lecture Notes in statistics*. B. Manly and L. McDonald (eds.), pp. 81–92. Springer-Verlag, Berlin.

Carey, J. R. (1982) Demography of the twospotted spider mite, *Tetranychus urticae* Koch. *Oecologia* 52:289–295.

Carey, J. R. (1983) Practical application of the stable age distribution: Analysis of a tetranychid mite (Acari: Tetranychidae) population outbreak. *Environ. Entomol.* 12:10–18.

Carey, J. R. (1984) Host-specific demographic studies of the Mediterranean fruit fly, *Ceratitis capitata. Ecol. Entomol.* 9:261–270.

Carey, J. R. (1986) Interrelations and applications of mathematical demography to selected problems in fruit fly management. In: *Pest Control: Operatons and Systems Analysis in Fruit Fly Management*. M. Mangel, J. Carey, and R. Plant (eds.), pp. 227–262. Springer-Verlag, Berlin.

Carey, J. R. (1991) Demographic approaches to specified problems in fruit fly management. In: S. Vijasegaran (ed.), *Proceedings 1st Int. Symp. Fruit Flies in Tropics*. Malasia Agric. Res. Dev. Inst., Kuala Lumpur, Malaysia.

Carey, J. R., and D. Krainacker (1988) Demographic analysis of tetranychid spider mite populations: Extensions of stable theory. *Exp. Appl. Acarol.* 4:191–210.

Carey, J. R., and R. I. Vargas (1985) Demographic analysis of insect mass rearing: A case study of three tephritids. *J. Econ. entomol.* 78:523–527.

Carey, J. R., T. T. Y. Wong, and M. M. Ramadan (1988). Demographic framework for parasitoid mass rearing: Case study of *Biosteres tryoni* (Hymenoptera: Braconidae), a larval parasitoid of tephritid fruit flies. *Theor. Popul. Biol.* 34:279–296.

Carnahan, B., H. A. Luther, and J. O. Wilkes (1969) *Applied Numerical Methods.* John Wiley & Sons, New York.

Davidson, G. (1954) Estimation of the survival-rate of Anopheline mosquitoes in nature. *Nature* 174:792–793.

Finney, D. J. (1964) *Probit Analysis.* Cambridge University Press, Cambridge.

Getz, Wyne M., and R. G. Haight (1989) *Population Harvesting.* Princeton University Press, Princeton.

Goodman, D. (1980) Demographic intervention for closely managed populations. In: *Conservation Biology: An Evolutionary-Ecological Perspective.* M. E. Soule and B. A. Wilcox (eds.). Sinauer Associates, Inc., Sanderland, Massachusetts, pp. 171–195.

Hairston, N. G., and S. Twombly (1985) Obtaining life table data from cohort analysis: A critique of current methods. *Limnol. Oceanogr.* 30:886–893.

Keyfitz, N. (1982) Choice of function for mortality analysis: Effective forecasting depends on a minimum parameter representation. *Theor. Popul. Biol.* 21:329–352.

Keyfitz, N. (1985) *Applied Mathematical Demography.* Springer-Verlag, New york.

King, B. H. (1987) Offspring sex ratios in parasitoid wasps. *Q. Rev. Biol.* 62:367–396.

Litchfield, J. T., and F. Wilcoxon (1949) A simplified method of evaluating dose-effect experiments. *J. Pharmacol Toxicol.* 96:99–113.

Manly, B. F. J. (1985) Further improvements to a method for analysing stage-frequency data. *Res. Popul. Ecol.* 27:325–332.

Manly, B., and L. Mcdonald, eds. (1989) *Estimation and Analysis of Insect Populations.* Lecture Notes in Statistics. Springer-Verlag, Berlin.

Mills, N. J. (1981) The estimation of mean duration from stage frequency data. *Oecologia* 51:206—211.

Morrison, J. K., R. M. Quinton, and H. Reinert (1968) The purpose and value of LD_{50} determinations. In: R. Gaylana and S. Goulding (eds.), *Modern Trends in Toxicology.* Butterworths, London.

Pearl, R., and L. J. Reed (1920) On the rate of growth of the population of the United States since 1790 and its mathematical representation. *Proc. nat. Acad. Sci. USA* 6:275–288.

Plant, R. (1986) The sterile insect technique: A theoretical perspective. In: M. Mangel, J. Carey and R. Plant (eds.). *Pest Control: Operations and Systems Analysis in Fruit Fly Management.* Springer-Verlag, Berlin, pp. 361–386.

Pontius, J. S., J. E. Boyer, Jr., and M. L. Deaton (1989) Estimation of stage transition time: Application to entomological studies. *Ann. Entomol. Soc. Am.* 82:135–148.

Appendix 1
A Preliminary Analysis of Mortality in 1.2 Million Medflies

BACKGROUND

Life tables have been published on a variety of organisms including a large number of insect species. Some of the more widely cited insect life tables include those for *Drosophila* by Pearl and Parker (1924), for the flour beetle (*Tribolium castaneum*) by Leslie and Park (1949), for the human louse (*Pediculus humanas*) by Evans and Smith (1952), and for the house fly (*Musca domestica*) by Rockstein and Lieberman (1959). The classic paper by Deevey (1947) contains life table information on several dozen vertebrate and invertebrate species, the paper by Promislow (1991) contains mortality data on 56 natural populations of mammals, and the books by Finch (1990) and Gavrilov and Gavrilova (1991) contain extensive bibliographies of life tables published on nonhuman organisms.

Unfortunately, the majority of published life tables for nonhuman species suffer three major shortcomings that restrict their potential for broader analysis. *First*, most are based on small initial numbers of individuals. This number is usually less than 50 individuals and only rarely have several thousand inviduals been used to construct a life table. The problem with small initial numbers is that, like all actuarial processes, the numbers decrease progressively with age due to attrition. Little can be learned about mortality patterns at older ages based on small absolute numbers of survivors. *Second*, few complete life tables are ever published because of space limitations in journals. Instead, the life table results are summarized and therefore the age patterns of the life table schedules are not retrievable. *Third*, the survival schedule is typically the only age-specific function published and this is usually as a graph. It is essentially impossible to extract reliable life table information from a graphed survival curve at ages when survival is less than around 5 to 10% (also see Finch and Ricklefs, 1991).

Because of the lack of extensive life table data on nonhuman species, in 1989 the U.S. National Institute on Aging funded a project titled "Oldest-Old Mortality for Mediterranean Fruit Flies" whose objective was to gather life table information on a minimum of one million medflies (see Barinaga, 1991). Particular emphasis was on mortality patterns at advanced ages. The project

APPENDIX

was part of a larger program entitled "Oldest-Old Mortality: Demographic Models and Analyses" originally administered through the Population Project of the Humphrey Institute at the University of Minnesota and later through the Center for Demographic Studies at Duke University. The medfly project was administered and coordinated at the University of California, Davis, with data collected on medflies at the Moscamed Program in Tapachula, Chiapas, Mexico—a large medfly mass-rearing facility capable of producing over 500 million medfly pupae weekly. The life tables contained in Appendices 2 and 3 are constructed from data gathered in 1991 at Moscamed on medflies held in 167 cages of approximately 7,200 flies each (see Carey et al., 1992).

The purpose of this appendix is to introduce methods for analysis of mortality data that are in the demographic literature but have not been introduced to the biology literature. The analysis presented here is intended to illustrate techniques as well as to highlight some of the more substantive aspects of the results. The overall analysis is presented in eight different parts: (1) life table functions; (2) male and female mortality rates; (3) mortality rate doubling time; (4) male-female crossovers for mortality, survival and life expectancy; (5) age groups responsible for relative differences in sex-specific life expectancies; (6) effects on life expectancies of proportional mortality modifications; (7) life table entropy; (8) estimation using computer "cohorts".

LIFE TABLES FUNCTIONS

The life tables for the mortality experience of 1,203,646 medflies were constructed for each sex; the results are presented in Appendix 2 for males and Appendix 3 for females. One male lived to 164 days and two females lived to 171 days. Graphs of four main life table functions based on life tables of all individuals (both sexes) combined are presented in Figure A1-1. The survival schedule (l_x) shows the fraction that survive to each age. A particularly striking aspect of the survival curve is the length of its tail—only around 10% of the original cohort lived beyond 30 days but some individuals survived for at least another 4 months. The d_x schedule shows that over 95% of all deaths occur in the first 6 weeks of life. The pattern of age-specific mortality schedule (q_x) is striking because it initially increases but then levels off and decreases at older ages. This pattern is contrary to the widely held belief that age-specific mortality continues to increase at advanced ages. Because of the deceleration of age-specific mortality, the schedule of expectations of life at age x(e_x) increases rather than decreases at advanced ages. These results cast doubt on several central concepts in demography, gerontology, and the biology of aging as outlined by Carey and co-workers (Carey et al., 1992) including: (1) that senescence can be characterized by the increase in age-specific mortality; (2) the basic pattern of mortality in nearly all species follows the same pattern at older ages (i.e. Gompertz exponential pattern); (3) species have specific life span limits.

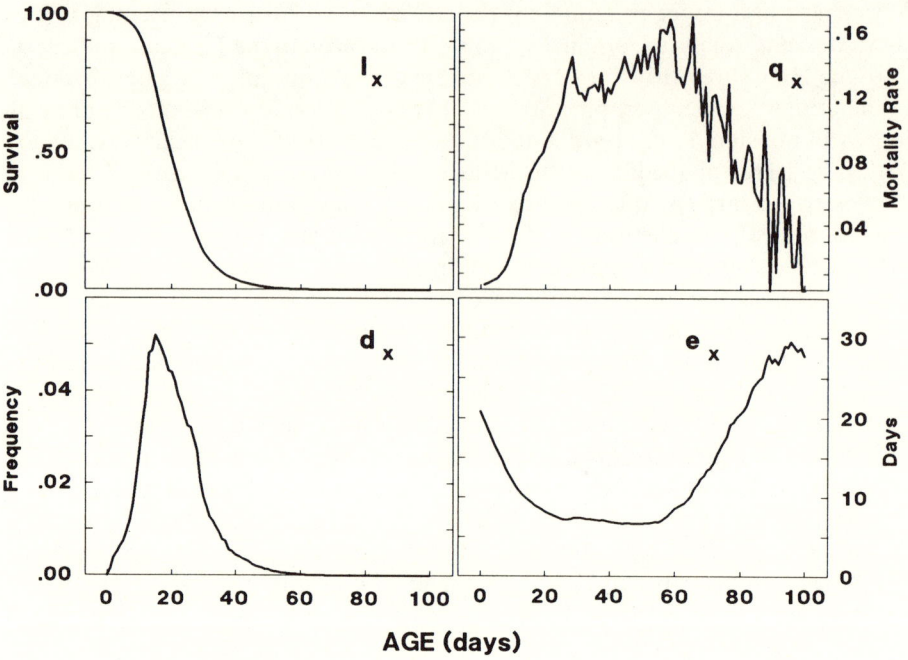

Figure A1-1. Life table functions for 1.2 million medflies using combined male and female data from Appendices 2 and 3.

MALE AND FEMALE AGE-SPECIFIC MORTALITY

Because of the binomial noise present in the sex-specific mortality schedules due to smaller numbers at the older ages, the sex-specific mortality rates were smoothed using the running geometric mean of the weekly mortality rate. This smoothed function is given as

$$q_x(smooth) = 1 - \left[\sqrt[7]{\prod_{y=x-3}^{x+3} (1 - q_y)} \right] \qquad \text{(A1-1)}$$

where q_x denotes the mortality rate at age x. The smoothed sex-specific mortality rates by age are presented in Figure A1-2. Three aspects of these patterns merit comment and will serve as the focus for many parts of the subsequent analyses: (1) at young ages the female mortality rates increase more rapidly than do male rates; (2) female mortality levels off at around 3 weeks of age whereas male mortality rates continue to increase until nearly 2 months of age; and (3) male age-specific mortality begins to decrease about 7 to 10 days before the female schedule, although at about 2 months the schedules decrease at the same rate. The subsequent analyses will focus on quantifying these differences in levels, patterns, and rates as well as gaining insights into the life table consequences of the differences.

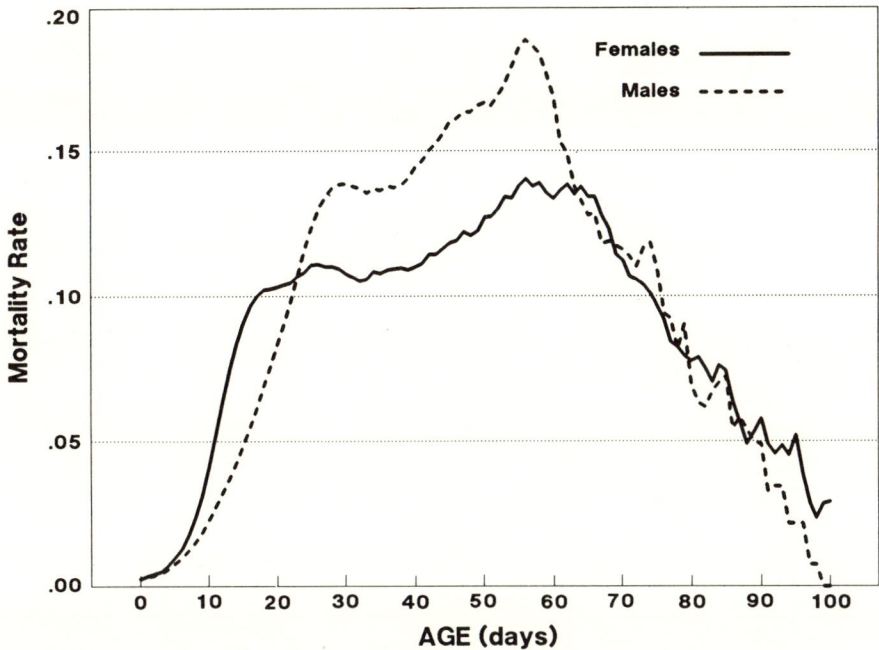

Figure A1-2. Smoothed age-specific mortality for male and female medflies from Appendices 2 and 3.

MORTALITY RATE DOUBLING TIME

Mortality rate doubling time (MRDT) stems from the exponential term, G, in the Gompertz equation $q_x = A_0 e^{Gx}$ (see Chapter 6) and denotes the time required for mortality rate to increase by 2-fold. This parameter is determined using the same formula for doubling times of any exponential process:

$$\text{MRDT} = \frac{\log_e 2}{G} \qquad (A1\text{-}2)$$

Taking natural logs of the age-specific mortality rates for each sex and estimating the slope of the best fitting straight line through these points from ages 0 through 14 days yields a 2-week exponential rate of change of $G = .2270$ for males and $G = .2936$ for females. Using the formula for MRDT given above yields a mortality rate doubling time of 3.1 days for males and 2.4 days for females. Or expressed another way, the mortality of young females increased by 16-fold in approximately the same time male mortality increased by 8-fold. An appendix in Finch (1990) contains estimates of MRDT's for a wide range of species.

CROSSOVERS

A useful technique for examining differences in life table functions between two populations is to examine their age ratios. The mortality sex ratio provides a perspective on the relative differences over the entire life course. For example, a ratio skewed towards males over all ages indicates that male mortality is greater than female mortality throughout the life couse. However, if the rates of change in age-specific mortality differ between sexes, that is, relatively greater at young ages and relatively less at older ages, a *mortality crossover* will occur. Manton and Stallard (1984) describe a mortality crossover (or convergence) in humans as an attribute of the relative rate of change and level of age-specific mortality rates in two population groups: one group is "advantaged" (i.e., lower relative mortality) and the other "disadvantaged" (higher relative mortality). The disadvantaged population must manifest age-specific mortality rates markedly higher than the advantaged population through middle age at which time the rates change. Age ratios of other functions including life expectancy and survival are also useful in identifying important relative patterns of the cummulative conseqences of mortality between the two sexes.

Three ratios of sex-specific life table functions were examined using the medfly life tables in Appendices 2 and 3 (superscripts m and f denote male and female rates, respectively):

$$\text{Mortality Sex Ratio} = \frac{q_x^m}{q_x^f} \qquad (A1\text{-}3a)$$

$$\text{Survival Sex Ratio} = \frac{l_x^m}{l_x^f} \qquad (A1\text{-}3b)$$

$$\text{Life Expectancy Sex Ratio} = \frac{e_x^m}{e_x^f} \qquad (A1\text{-}3c)$$

The survival sex ratio is important because it provides insights into the relative *retrospective effects* of the age-specific mortality schedules—the cumulative effects of daily mortality prior to age x. The life expectancy sex ratio sheds light on the relative *prospective effects* of the age-specific mortality schedules—the cumulative effects of daily mortality after age x through the last day of life.

The mortality sex ratio for the medflies was less than unity between ages 0 and 20 days indicating that male rates were less than female rates (Fig. A1-3). The maximal proportional differences in these sex-specific rates occurred near age 16 days. The first mortality crossover occurred around 2 weeks later (at 30 days) and a second mortality crossover occurred at around 70 days after which time the ratios fluctuated around 1.0. Manton and Stallard (1984) note that crossovers in mortality occur due to: (1) differences in rates of aging at the individual level; and (2) demographic selection where individuals with high mortality are selected out early for one population and therefore the more robust individuals (those with inherently lower mortality risks) survive to the

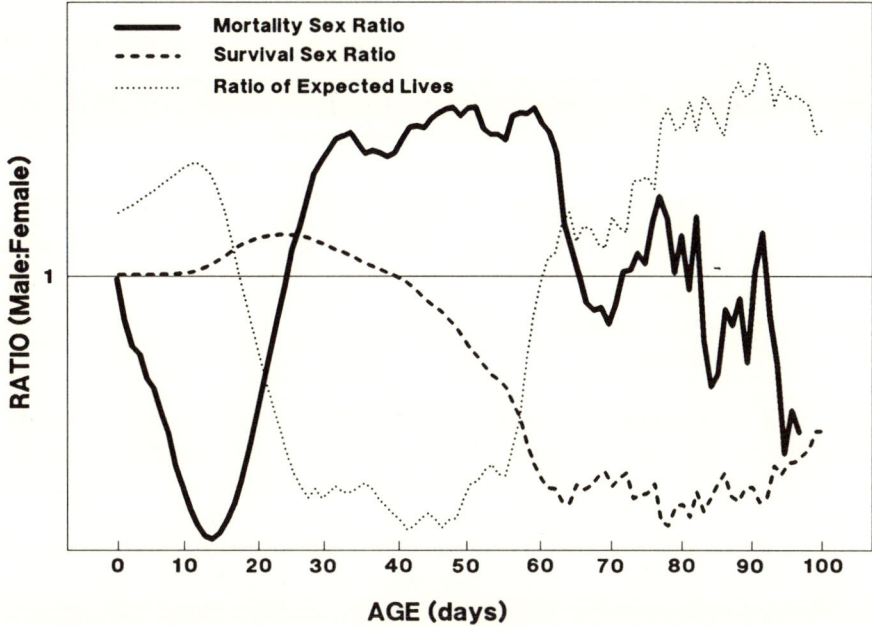

Figure A1-3. Mortality sex ratios, survival sex ratios, and ratios of sex-specific expectations of remaining life using medfly life table data in Appendices 2 and 3.

older ages. This is also referred to by Hobcraft et al. (1982) as the "cohort-inversion model" which is based on the concept that cohorts experiencing particularly hard or good times early in life will respond inversely later in life. The mortality sex ratio provides insights into the magnitude of mortality differences between the two sexes at each age but they do not quantify the extent to which these differences contribute to differences in life expectancy or survival or the sensitivity of life expectancy to changes in mortality at selected ages.

Two survival schedules that differ due to mortality differences wil *converge* if a mortality crossover occurs and if this crossover persists for a sufficient time to offset survival differentials that existed prior to the crossover. If the mortality differentials continue to persist then the survival schedules will eventually crossover. Survival crossovers late in life may suggest that differences between mortality in two populations are not due simply to differences in age-specific rates of aging that eventually are offsetting. Rather, they may ultimately reflect lifetime differences in rates of aging between the two populations. Note that in Figure A1-3 survival crossover occurs at around 40 days shifting from favoring males to favoring females. This female advantage persists until the last two individuals in the cohort die—both females.

Crossovers of remaining life expectancies can only occur if mortality crossovers occur at a future age rather than at a previous age as is the requirement for survival crossovers. There are two life expectancy crossovers for the male and female medflies that divide the life course into three periods: (1) prior to age 20 days male life expectancy exceeds female life expectancy; (2) from age 20 to 60 female life expectancy exceeds that of the males; and (3) after age 60 males' life expectancy again exceeds that of females. It is noteworthy that males are in the minority after age 60, as indicated by the survival ratios, yet male life expectancy after this age is greater. This particular combination of ratios indicates that females are more likely to survive to old age but, on average, old males live longer.

AGE GROUPS RESPONSIBLE FOR SEX DIFFERENCES IN LIFE EXPECTANCIES

Life expectancies between male and female medflies differed by 2.55 days: 19.58 days for females and 22.13 days for males. The question of which age groups account for these differences can be addressed using the formula for decomposing these differences between two cohorts (Anon, 1983). Let $_n\Delta_x$ denote the change in expectation of life between male and females from age x to age $x + n$, l_x and e_x denote the life table functions survival to age x and expectation of life at age x, respectively, and superscripts m and f denote male and female, respectively. Then

$$_n\Delta_x = A - B \qquad (A1\text{-}4a)$$

where

$$A = (e_x^f - e_x^m)\frac{l_x^f + l_x^m}{2} \qquad (A1\text{-}4b)$$

and

$$B = (e_{x+n}^f - e_{x+n}^m)\frac{l_{x+n}^f + l_{x+n}^m}{2} \qquad (A1\text{-}4c)$$

Analysis of the intervals 0 to 10, 11 to 20, 21 to 30, and 30 to the last day of life using the appropriate sex-specific values of survival and expectation of life yielded the results for differences in expectation of life between males and females given in Table A1-1.

Table A1-1. Measure of the Contribution of Sex Mortality Differences at Various Ages to the Gap in Life Expectancy at Eclosion

Age Group (days)	Difference due Age Group	Difference in Percent	Difference Favours
0–10	−0.30	−12%	males
11–20	−2.60	−102%	males
21–30	0.08	3%	females
30+	0.27	11%	females
Total	−2.55	−100%	

A positive number indicates a contribution in favor of females and a negative number indicates a contribution in favor of males. Note that all positive numbers occur for the two age groups greater than 20 days indicating contributions in favor of females and all negative (i.e., male) contributions occur for the first two age groups: 0 to 10 and 11 to 20. This reflects the difference in sex-specific mortality levels and rates of change during the first 20 days. Indeed, these differences account for nearly the entire 2.55 day difference in life expectancies between the sexes. Note also that the small differences in age group 21 to 30 reflects the fact that it is during this period that a mortality crossover occurs—contributions toward males early in this period are offset by the contributions by females later in the period. One of the main implications of this result is that it focuses attention on the clear need to understand the underlying biological differences in the two sexes during the first 20 days.

EFFECT OF PROPORTIONAL MORTALITY DIFFERENCES ON LIFE EXPECTANCIES

Male and female medfly life expectancies at eclosion differ by 13%. This difference raises an important question that has been addressed by a number of demographers including Pollard (1982), Keyfitz (1985), and Vaupel (1986): "How much do the mortality rates differ that ultimately determine the life expectancy differences?"

This question can be addressed by letting $\hat{e}_0^m$ denote a new male expectation of life at age 0 computed using a modified age-specific mortality schedule $\hat{q}_x^m$ where

$$\hat{q}_x^m = (1 + \delta)q_x^m \quad (A1\text{-}5)$$

and q_x^m denotes the original male age-specific mortality rate at age x. Likewise let $\hat{e}_0^f$ denote a new female expectation of life at age 0 computed using a modified age-specific mortality schedule $\hat{q}_x^f$ where

$$\hat{q}_x^f = (1 + \delta)q_x^f \quad (A1\text{-}6)$$

and q_x^f denotes the original female age-specific mortality rate at age x. Results of computing sex-specific life expectancies using different δ-values to modify age-specific mortality are given in Table A1-2.

The interpretation for the case when $\delta = -0.2$ indicates that if age-specific mortality schedule is multiplied by a factor that reduces the rates uniformly across all age classes by 20% (i.e. 80% of the original) then the expectation of life of a newly-enclosed male medfly increases by 10% from 22.1 days to 24.3 days, and for a female medfly the expectation of life increases by 12% from 19.6 days to 22.0 days. This analysis shows that: (1) a proportional change in age-specific mortality of a given schedule does not have a proportional effect on expectation of life; (2) equivalent proportional changes in two mortality schedules that have different age patterns will not result in identical relative changes in the respective cohort life expectancies. For example, as shown in

Table A1-2. Effect of a Uniform Change in Age-Specific Mortality by a Factor of $(1+\delta)$ on Male and Female Life Expectancies

Changes in Mortality Schedule	δ-value	Male $\hat{e}_0^m$	Female $\hat{e}_0^f$
Decrease by 20%	−0.2	24.3 days	22.0 days
No change	.0	22.1*	19.6*
Increase by 20%	.2	20.6	18.0
Increase by 40%	.4	19.4	16.8

Numbers with asterisk (*) indicated expectation of life resulting from the unaltered mortality schedules

Table A1-2 the fact that male medflies have a 13% longer expected life than female medflies is equivalent to males having a 20% lower mortality than females with respect to modifications of the female schedule. However, this 13% differential in expected life is also equivalent to females having a 40% higher mortality than males using the male schedule as reference.

The different factors by which mortality must be changed for each sex to produce equivalent life expectancies is due to differences in age patterns of mortality: the female schedule rises more rapidly at young ages and levels off earlier than the male mortality schedule. Therefore, a proportional increase of both schedules will widen the *absolute* mortality differences at young ages. The importance of this analysis is that it quantifies the effect on expectation of life differences in mortality *patterns*.

LIFE TABLE ENTROPY

Recall from Chapter 2 that life table entropy, H, is the sum of the products over all ages of the life table functions e_x and d_x divided by expectation of life at age 0, e_0. That is

$$H = \frac{\sum_{x=0}^{\omega} d_x e_x}{e_0} \qquad (A1\text{-}7)$$

Using the appropriate schedules from Appendices 2 and 3, the entropy-value for medfly males was computed as $H = 0.393$ and for females as $H = 0.477$. These values have two implications. First, H is a summary of the degree of concavity of the survival schedule—$H = 0$ indicates extreme concavity (i.e., square) and $H = 1$ indicated convexity associated with a geometric decrease in survival with age. Thus the male medfly survival schedule is more concave than is the female survival schedule. *Second*, a 1% change in the mortality schedule for males will change expectation of life by .393% whereas a 1% change in the mortality schedule for females will change female expectation of life by .477%. Therefore, a small change in female mortality will have a greater impact on female expectation of life than a small change in male mortality will have on male expectation of life, *ceteris peribus*.

Table A1-3. Value of the Entropy Numerator Product $e_x d_x$ Over 10-day Age Intervals for Male and Female Medflies. Results Give the Average Days of Future Life that Are Lost by the Observed Deaths

Age Interval (x to x + 10)	Numerator of H $\sum_{y=x}^{x+10} d_y e_y$	
	Males	Females
0–10	1.45	1.55
10–20	3.99	5.22
20–30	3.06	2.35
30–40	.84	.71

The effect of uniform change in mortality across all ages on expectation of life are different at each age, depending upon the product of e_x and d_x. The age with the highest $e_x d_x$ value was 17 days for males and 13 days for females. The sum of the products of $e_x d_x$ summed over 10-day age groups is contained in Table A1-3. These results show that 5.22 days are lost by females due to deaths in the age interval 10 to 20 days and that this is 2- to 5-fold greater than the effect of a change of mortality in any of the other 10-day intervals. In contrast, the effect on male mortality of changes in mortality over these corresponding age groups is less than that for females from 10 to 20 days but greater at older ages.

These results are important because they show that the magnitude of the effects of changes in mortality rates on life expectancy differs substantially between the sexes. Entropy sheds light on the sensitivity of expectation of life to small *changes* in mortality rates among the different age groups.

THE FORCE OF MORTALITY AND OLDEST-OLD

The force of mortality at age x, denoted $\mu(x)$, is the instantaneous mortality rate representing the limiting value of the age-specific mortality rate when the age interval to which the rate refers becomes infinitesimally short (Pressat, 1985). It is given as

$$\mu(x) = -\frac{dl(x)}{l(x)dx} \tag{A1-8}$$

Also known as the instantaneous death rate and hazard rate, $\mu(x)$ is used extensively in gerontology, actuarial mathematics (e.g. Bowers et al. 1986), demography, and the analysis of life spans (e.g., Gavrilov and Gavrilova, 1991) because it has many useful properties, the most important of which is that its value does not depend on the length of the age interval. In contrast, the

Table A1-4. Force of Mortality for Male and Female Medflies at Five Advanced Ages

Age (days)	Force of Mortality	
	Males	Females
100	.0054	.0340
110	.0251	.0279
120	.0154	.0388
130	.0288	.0547
140	.0405	.0318

magnitude of age-specific mortality depends on the length of the age interval. This property of the force of mortality is particularly important for testing the hypotheses concerned with upper life span limits. Gavrilov and Gavrilova (1991) note that if a species limit to life span exists, at the maximal age the number of survivors should become zero, the probability of death should equal unity, and the force of mortality at this age should tend towards plus infinity. As they state, "Thus in order to test the hypothesis in question, it is sufficient to investigate the characteristics of the life span distribution in the maximal age region and check whether at these ages there really is a catastropic increase in the force of mortality."

The formula proposed by Sacher (1966) used to estimate the force of mortality at age x is given as

$$\mu(x) = \frac{1}{2n} \ln_e \left(\frac{l(x-n)}{l(x+n)} \right) \tag{A1-9}$$

One of the objectives of the medfly study was to examine the age pattern of the force of mortality at advanced ages. If a species-specific life span exists, the force of mortality should exibit catastrophic increase at the advanced ages as individuals approach their maximal age. The force of mortality was computed using (A1-9) for both sexes at ages 100 days or greater in 10 day increments with $n = 5$. The results are given in Table A1-4.

It is clear from Table A1-4 that the force of mortality at advanced ages for both sexes does not show a catastropic increase at any of the advanced ages but rather varies between 0.5% and 4% for males and between 2% and 5.5% for females. It follows from this result that a maximal age with special properties does not exist. As Gavrilov and Gavrilova (1991) note, this fact refutes the idea that "locked within the code of the genetic material are instructions that specify the age beyond which a species cannot live given even the most favourable conditions" (The New Encyclopaedia Britannica, 1989, Vol. 20, p. 471).

ESTIMATION

The logistics required for monitoring mortality patterns in cohorts ranging in size from several thousand to a million or more are beyond the experimental

Table A1-5 Results of Computer-Generated Medfly Cohort Simulation Study. Actual Expectation of Life at Age 0 Was $e_0 = 20.8$ days and the Actual Oldest Fly Recorded in the Study was 171 Day

Cohort Size (100 reps)	Expectation of Life				Oldest Fly			
	Min	Max	Ave	SD	Min	Max	Ave	SD
25	15.4	26.7	20.8	2.05	26	80	42.1	8.57
100	18.6	22.9	20.6	.89	37	88	54.3	9.99
1,000	19.9	21.2	20.7	.26	53	163	70.0	15.08

capabilities of many biologists. However, it is still possible to examine questions of sampling and estimation based on computer-generated cohorts. As an example, longevities of individual "flies" were created by subjecting an "individual" to the smoothed medfly mortality schedule for both sexes combined. Hypothetical individuals "died" at the first age when a random number between 0 and 1 was less than or equal to the observed mortality rate at that age. A total of 100 "cohorts" were created in this way for each of the following cohort sizes: 25, 200, and 1,000 flies per cohort. This resulted in 100 groups of 25 flies each (i.e., total of 2,500 flies), 100 groups of 100 flies each (i.e., total of 10,000 flies), and 100 groups of 1,000 flies each (i.e., 100,000 flies). The mortality experience of these computer cohorts were used to produce life tables for each cohort, the results of which are summarized in Table A1-5.

These results of this simulation merit two comments. *First*, larger numbers of flies per cohort produced the smallest variation in life expectancies. This was anticipated. For example, the difference in life expectancies between cohorts with 25 flies was over 11 days. In contrast, the difference in life expectancies was less than 1.5 days for the 100 replicates of 1,000 flies per cohort. For all three cohort sizes, the average of all 100 cohort life expectancies was close to the actual observed value of 20.8 days. *Second*, the range, average, and standard deviation for the oldest flies in the 100 sets of different-sized cohorts all increased with cohort size. For example, one 25-fly cohort died out as early as 26-days (oldest fly) and one died out as late as 80 days—a range of 54 days. However, one 1000-fly cohort died out as early as 53 days and one as late as 163 days—a range of 110 days. Thus the oldest age attained in some 1000-fly cohorts was less than the oldest age attained in some 25-fly cohorts due to chance alone. This result underscores the precariousness of the measurement of "maximal age." Cohort size and chance play enormous roles in determining this parameter.

Age-specific mortality rates were determined for computer-generated cohorts ranging in size from 25 to 100,000 flies. The original smoothed age-specific mortality schedule for the 1.2 million medflies was the probability function from which all computer cohorts and, in turn, all other mortality schedules were generated using the random number simulation technique described earlier. It is evident from the results given in Fig. A1-4 that the variability of mortality rates at older ages as well as the age of the oldest fly is heavily dependent upon initial cohort size. This figure also shows that the contours

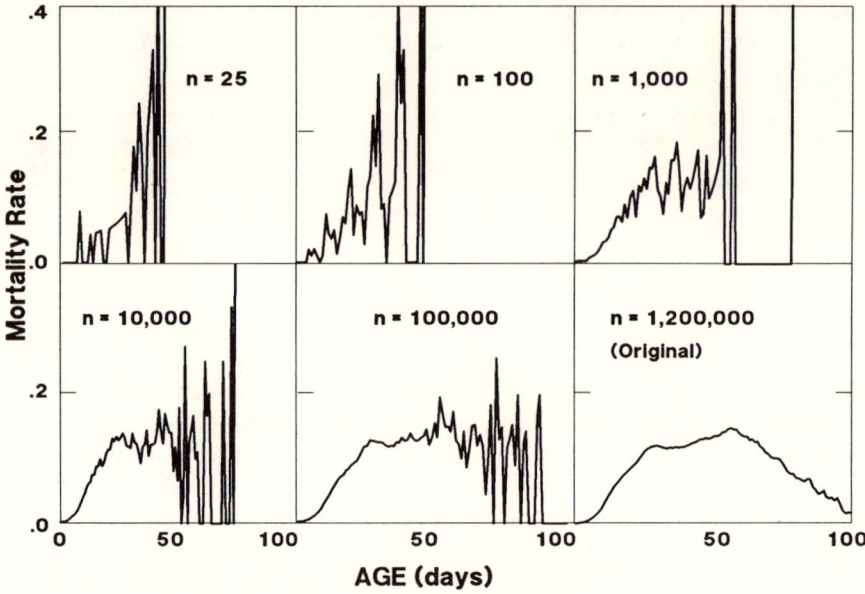

Figure A1-4. Results of various cohort sizes for estimating age-specific mortality patterns in the medfly. The cohorts were generated on the computer using the schedule for 1.2 million medflies (lower right graph). See text.

of the overall mortality pattern become progressively smoother and more precise with age as cohort size increases. The two cohorts of n = 25 and n = 100 flies died out in less than 50 days and the age trends in mortality rates for both cohorts appear to increase monotonically. However, the mortality patterns in these two cohorts reveal virtually nothing about the actual underlying function; none of the age contours of mortality are pronounced and even the broad outline of mortality at young ages is blurred. And the putative maximal age in the actual cohort was over 300% greater than the oldest fly in either of these two cohorts. About the only aspect of the mortality schedules from these two small cohorts that is consistent with the actual one is their sign—mortality rates in both increase at young ages. This result is disturbing since, as noted previously, 25 to 100 individuals represent the most common range of initial cohort sizes upon which the vast majority of nonhuman life tables are based.

The mortality pattern resulting from the 1,000-fly cohort is definitive for the first 20–30 days but only suggestive with respect to the leveling off. This hint of leveling off is similar to that reported by Rockstein and Lieberman (1959) for age-specific mortality in 4,000 male houseflies. At n = 10,000 flies the deceleration and leveling off is more pronounced but only at n = 100,000 does the distinctive profile of the plateau emerge. The mortality decrease in the right tail is masked by binomial noise in even the largest computer-generated cohort of 100,000 individuals. Of course this result opens up questions regarding medfly mortality patterns that lie beyond age 171 days which would require cohorts of perhaps 10 million or a billion flies to investigate. Indeed, the problem is open ended, literally.

The practical problem at issue here is not concerned with estimating the specifics of the medfly mortality schedule, *per se*, or whether a maximal age exists. Rather the general issue has to do with estimating *any* age-specific mortality schedule. Observed mortality patterns may have more to do with the problems inherent in the use of small cohort numbers than with the nature of the phenomenon.

BIBLIOGRAPHY

Anon, (1983) Patterns of sex differentials in mortality in less developed countries. In: *Sex Differential in Mortality: Trends, Determinations and Consequences*. A. D. Lopez and L. T. Ruzicka (eds). Department of Demography, Australian National University Press, Canberra.

Barinaga, M. (1991) How long is the human life-span? *Science* 254: 936–938.

Bowers, N. L., Jr., H. U. Gerber, J. C. Hickman, D. A. Jones and C. J. Nesbitt (1986) *Actuarial Mathematics*. The Society of Actuaries, Itasca, Illinois.

Carey, J. R., P. Liedo, D. Orozco and J. W. Vaupel (1992) Slowing of mortality rates at older ages in large medfly cohorts. *Science* (in press)

Coale, A. J., and E. E. Kisker (1986) Mortality crossovers: reality or bad data? *Popul. Stud.* 40: 389–401.

Finch, C. E. (1990) *Longevity, Senescence, and the Genome*. The University of Chicago Press, Chicago.

Finch, C. E., and R. E. Ricklefs (1991) Age structure of populations. *Science* 254: 779.

Gavrilov, L. A., and N. S. Gavrilova (1991) *The Biology of Life Span: A Quantitative Approach*. Harwood Academic Publishers, Chur, Switzerland.

Hazzard, DeWitt, G., H. R. Warner, and Caleb E. Finch (1991) National Institute on Aging, NIH, Workshop on Alternative Animal Models for Research on Aging. *Expt. Gerontol.* 26: 411–439

Hobcraft, J. J. Menden, and S. Preston, (1982) Age, period and cohort effects in demography: A review. *Popul. Index.* 48: 4–43

Keyfitz, N. (1985) *Applied Mathematical Demography*. (2nd Ed.), Springer-Verlag, New York.

Manton, K., and E. Stallard (1984) *Recent Trends in Mortality Analysis*. Academic Press, New York.

Pressat, R. (1985) *The Dictionary of Demography*. Bell and Bain, Ltd. Glasgow.

Promislow, D. E. L. (1991) Senescence in natural populations of mammals: A comparative study. *Evolution* 45: 1869–1887.

Rockstein, M., and H. M. Lieberman. (1959) A life table for the common house fly, *Musca domestica. Gerontologia* 3: 23–36.

Vaupel, J. W. (1986) How change in age-specific mortality affects life expectancy. *Popul. Stud.* 40: 147–157.

Vaupel, J. W., and A. I. Yashin (1987) Repeated resuscitation: How lifesaving alters life tables. *Demography* 24: 123–135.

Appendix 2
Life Table for 598,118 Male Medflies

Table A2-1. Life Table for 598,118 Male Medflies Reared in a Total of 167 Cages at the Moscamed Mass Rearing Facility in Tapachula, Chiapas, Mexico. Data Were Gathered for the Project Titled "Oldest-Old Mortality in the Mediterranean Fruit Fly" Funded by the National Institute on Aging (Unpublished Data from J. R. Carey, P. Liedo, D. Orozco, and J. W. Vaupel)

Age Class	Number Dying	Number Alive	Living at Age x	Fraction Surviving from x to x+1	Fraction Dying from x to x+1	Dying in Interval x to x+1	Days Lived in Interval	Days Lived Beyond Age x	Expectation of Life
x	D_x	N_x	l_x	p_x	q_x	d_x	L_x	T_x	e_x
(1)	(2)	(3)	(4)	(5)	(6)	(7)	(8)	(9)	(10)
0	0	598118	1.00000	1.00000	.00000	.0000	1.00000	22.13	22.13
1	898	598118	1.00000	.99850	.00150	.0015	.99925	21.13	21.13
2	2468	597220	.99850	.99587	.00413	.0041	.99644	20.13	20.16
3	2946	594752	.99437	.99505	.00495	.0049	.99191	19.13	19.24
4	3383	591806	.98945	.99428	.00572	.0057	.98662	18.14	18.33
5	3875	588423	.98379	.99341	.00659	.0065	.98055	17.15	17.43
6	5166	584548	.97731	.99116	.00884	.0086	.97299	16.17	16.55
7	5873	579382	.96868	.98986	.01014	.0098	.96377	15.20	15.69
8	8145	573509	.95886	.98580	.01420	.0136	.95205	14.23	14.85
9	9810	565364	.94524	.98265	.01735	.0164	.93704	13.28	14.05
10	13058	555554	.92884	.97650	.02350	.0218	.91792	12.35	13.29
11	13892	542496	.90700	.97439	.02561	.0232	.89539	11.43	12.60
12	16388	528604	.88378	.96900	.03100	.0274	.87008	10.53	11.92
13	19817	512216	.85638	.96131	.03869	.0331	.83981	9.66	11.28
14	20138	492399	.82325	.95910	.04090	.0337	.80641	8.82	10.72
15	22698	472261	.78958	.95194	.04806	.0379	.77060	8.02	10.15
16	23954	449563	.75163	.94672	.05328	.0400	.73160	7.25	9.64
17	26601	425609	.71158	.93750	.06250	.0445	.68934	6.51	9.15
18	27341	399008	.66711	.93148	.06852	.0457	.64425	5.82	8.73
19	27657	371667	.62139	.92559	.07441	.0462	.59827	5.18	8.34
20	28915	344010	.57515	.91595	.08405	.0483	.55098	4.58	7.97
21	28984	315095	.52681	.90802	.09198	.0485	.50258	4.03	7.65
22	27935	286111	.47835	.90236	.09764	.0467	.45500	3.53	7.38
23	26990	258176	.43165	.89546	.10454	.0451	.40908	3.07	7.12
24	25993	231186	.38652	.88757	.11243	.0435	.36479	2.66	6.89
25	25023	205193	.34306	.87805	.12195	.0418	.32215	2.30	6.70
26	24855	180170	.30123	.86205	.13795	.0416	.28045	1.98	6.56
27	22836	155315	.25967	.85297	.14703	.0382	.24058	1.70	6.54
28	19303	132479	.22149	.85429	.14571	.0323	.20536	1.46	6.58

APPENDIX

Table A2-1. *(Continued)*

Age Class	Number Dying	Number Alive	Living at Age x	Fraction Surviving from x to x+1	Fraction Dying from x to x+1	Dying in Interval x to x+1	Days Lived in Interval	Days Lived Beyond Age x	Expectation of Life
x	D_x	N_x	l_x	p_x	q_x	d_x	L_x	T_x	e_x
(1)	(2)	(3)	(4)	(5)	(6)	(7)	(8)	(9)	(10)
29	16248	113176	.18922	.85644	.14356	.0272	.17564	1.25	6.61
30	13225	96928	.16205	.86356	.13644	.0221	.15100	1.08	6.64
31	11067	83703	.13994	.86778	.13222	.0185	.13069	.92	6.61
32	9486	72636	.12144	.86940	.13060	.0159	.11351	.79	6.54
33	8816	63150	.10558	.86040	.13960	.0147	.09821	.68	6.44
34	7686	54334	.09084	.85854	.14146	.0129	.08442	.58	6.41
35	6484	46648	.07799	.86100	.13900	.0108	.07257	.50	6.38
36	5334	40164	.06715	.86719	.13281	.0089	.06269	.43	6.33
37	5100	34830	.05823	.85357	.14643	.0085	.05397	.36	6.22
38	3829	29730	.04971	.87121	.12879	.0064	.04651	.31	6.21
39	3631	25901	.04330	.85981	.14019	.0061	.04027	.26	6.05
40	3029	22270	.03723	.86399	.13601	.0051	.03470	.22	5.95
41	2873	19241	.03217	.85068	.14932	.0048	.02977	.19	5.81
42	2530	16368	.02737	.84543	.15457	.0042	.02525	.16	5.74
43	2208	13838	.02314	.84044	.15956	.0037	.02129	.13	5.70
44	1878	11630	.01944	.83852	.16148	.0031	.01787	.11	5.69
45	1495	9752	.01630	.84670	.15330	.0025	.01505	.09	5.69
46	1300	8257	.01380	.84256	.15744	.0022	.01272	.08	5.63
47	1113	6957	.01163	.84002	.15998	.0019	.01070	.07	5.59
48	1036	5844	.00977	.82272	.17728	.0017	.00890	.05	5.56
49	791	4808	.00804	.83548	.16452	.0013	.00738	.05	5.65
50	698	4017	.00672	.82624	.17376	.0012	.00613	.04	5.66
51	540	3319	.00555	.83730	.16270	.0009	.00510	.03	5.75
52	469	2779	.00465	.82123	.16877	.0008	.00425	.03	5.77
53	378	2310	.00386	.83636	.16364	.0006	.00355	.02	5.84
54	295	1932	.00323	.84731	.15269	.0005	.00298	.02	5.89
55	326	1637	.00274	.80086	.19914	.0005	.00246	.02	5.86
56	254	1311	.00219	.80625	.19375	.0004	.00198	.01	6.19
57	226	1057	.00177	.78619	.21381	.0004	.00158	.01	6.55
58	173	831	.00139	.79182	.20818	.0003	.00124	.01	7.20
59	125	658	.00110	.81003	.18997	.0002	.00100	.01	7.96
60	80	533	.00089	.84991	.15009	.0001	.00082	.01	8.71
61	60	453	.00076	.86755	.13245	.0001	.00071	.01	9.16
62	59	393	.00066	.84987	.15013	.0001	.00061	.01	9.48
63	46	334	.00056	.86228	.13772	.0001	.00052	.01	10.07
64	30	288	.00048	.89583	.10417	.0001	.00046	.01	10.60
65	46	258	.00043	.82171	.17829	.0001	.00039	.00	10.78
66	24	212	.00035	.88679	.11321	.0000	.00033	.00	12.00
67	22	188	.00031	.88298	.11702	.0000	.00030	.00	12.47
68	16	166	.00028	.90361	.09639	.0000	.00026	.00	13.06
69	23	150	.00025	.84667	.15333	.0000	.00023	.00	13.40
70	8	127	.00021	.93701	.06299	.0000	.00021	.00	14.74
71	13	119	.00020	.89076	.10924	.0000	.00019	.00	14.69
72	18	106	.00018	.83019	.16981	.0000	.00016	.00	15.43
73	9	88	.00015	.89773	.10227	.0000	.00014	.00	17.49
74	8	79	.00013	.89873	.10127	.0000	.00013	.00	18.42
75	5	71	.00012	.92958	.07042	.0000	.00011	.00	19.44

Table A2-1. (Continued)

Age Class	Number Dying	Number Alive	Living at Age x	Fraction Surviving from x to x+1	Fraction Dying from x to x+1	Dying in Interval x to x+1	Days Lived in Interval	Days Lived Beyond Age x	Expectation of Life
x	D_x	N_x	l_x	p_x	q_x	d_x	L_x	T_x	e_x
(1)	(2)	(3)	(4)	(5)	(6)	(7)	(8)	(9)	(10)
76	13	66	.00011	.80303	.19697	.0000	.00010	.00	18.88
77	4	53	.00009	.92453	.07547	.0000	.00009	.00	23.63
78	2	49	.00008	.95918	.04082	.0000	.00008	.00	24.52
79	3	47	.00008	.93617	.06383	.0000	.00008	.00	24.54
80	4	44	.00007	.90909	.09091	.0000	.00007	.00	25.18
81	1	40	.00007	.97500	.02500	.0000	.00007	.00	26.65
82	5	39	.00007	.87179	.12821	.0000	.00006	.00	26.32
83	2	34	.00006	.94118	.05882	.0000	.00006	.00	29.12
84	1	32	.00005	.96875	.03125	.0000	.00005	.00	29.91
85	1	31	.00005	.96774	.03226	.0000	.00005	.00	29.85
86	3	30	.00005	.90000	.10000	.0000	.00005	.00	29.83
87	3	27	.00005	.88889	.11111	.0000	.00004	.00	32.09
88	1	24	.00004	.95833	.04167	.0000	.00004	.00	35.04
89	0	23	.00004	1.00000	.00000	.0000	.00004	.00	35.54
90	2	23	.00004	.91304	.08696	.0000	.00004	.00	34.54
91	0	21	.00004	1.00000	.00000	.0000	.00004	.00	36.79
92	0	21	.00004	1.00000	.00000	.0000	.00004	.00	35.79
93	2	21	.00004	.90476	.09524	.0000	.00003	.00	34.79
94	0	19	.00003	1.00000	.00000	.0000	.00003	.00	37.39
95	1	19	.00003	.94737	.05263	.0000	.00003	.00	36.39
96	0	18	.00003	1.00000	.00000	.0000	.00003	.00	37.39
97	0	18	.00003	1.00000	.00000	.0000	.00003	.00	36.39
98	0	18	.00003	1.00000	.00000	.0000	.00003	.00	35.39
99	0	18	.00003	1.00000	.00000	.0000	.00003	.00	34.39
100	0	18	.00003	1.00000	.00000	.0000	.00003	.00	33.39
101	0	18	.00003	1.00000	.00000	.0000	.00003	.00	32.39
102	0	18	.00003	1.00000	.00000	.0000	.00003	.00	31.39
103	0	18	.00003	1.00000	.00000	.0000	.00003	.00	30.39
104	0	18	.00003	1.00000	.00000	.0000	.00003	.00	29.39
105	0	18	.00003	1.00000	.00000	.0000	.00003	.00	28.39
106	0	18	.00003	1.00000	.00000	.0000	.00003	.00	27.39
107	0	18	.00003	1.00000	.00000	.0000	.00003	.00	26.39
108	0	18	.00003	1.00000	.00000	.0000	.00003	.00	25.39
109	0	18	.00003	1.00000	.00000	.0000	.00003	.00	24.39
110	1	18	.00003	.94444	.05556	.0000	.00003	.00	23.39
111	1	17	.00003	.94118	.05882	.0000	.00003	.00	23.74
112	0	16	.00003	1.00000	.00000	.0000	.00003	.00	24.19
113	1	16	.00003	.93750	.06250	.0000	.00003	.00	23.19
114	1	15	.00003	.93333	.06667	.0000	.00002	.00	23.70
115	1	14	.00002	.92857	.07143	.0000	.00002	.00	24.36
116	0	13	.00002	1.00000	.00000	.0000	.00002	.00	25.19
117	0	13	.00002	1.00000	.00000	.0000	.00002	.00	24.19
118	0	13	.00002	1.00000	.00000	.0000	.00002	.00	23.19
119	0	13	.00002	1.00000	.00000	.0000	.00002	.00	22.19
120	0	13	.00002	1.00000	.00000	.0000	.00002	.00	21.19
121	0	13	.00002	1.00000	.00000	.0000	.00002	.00	20.19
122	1	13	.00002	.92308	.07692	.0000	.00002	.00	19.19

APPENDIX

Table A2-1. (Continued)

Age Class	Number Dying	Number Alive	Living at Age x	Fraction Surviving from x to x+1	Fraction Dying from x to x+1	Dying in Interval x to x+1	Days Lived in Interval	Days Lived Beyond Age x	Expectation of Life
x	D_x	N_x	l_x	p_x	q_x	d_x	L_x	T_x	e_x
(1)	(2)	(3)	(4)	(5)	(6)	(7)	(8)	(9)	(10)
123	0	12	.00002	1.00000	.00000	.0000	.00002	.00	19.75
124	0	12	.00002	1.00000	.00000	.0000	.00002	.00	18.75
125	0	12	.00002	1.00000	.00000	.0000	.00002	.00	17.75
126	2	12	.00002	.83333	.16667	.0000	.00002	.00	16.75
127	0	10	.00002	1.00000	.00000	.0000	.00002	.00	19.00
128	0	10	.00002	1.00000	.00000	.0000	.00002	.00	18.00
129	0	10	.00002	1.00000	.00000	.0000	.00002	.00	17.00
130	0	10	.00002	1.00000	.00000	.0000	.00002	.00	16.00
131	0	10	.00002	1.00000	.00000	.0000	.00002	.00	15.00
132	1	10	.00002	.90000	.10000	.0000	.00002	.00	14.00
133	0	9	.00002	1.00000	.00000	.0000	.00002	.00	14.50
134	0	9	.00002	1.00000	.00000	.0000	.00002	.00	13.50
135	0	9	.00002	1.00000	.00000	.0000	.00002	.00	12.50
136	1	9	.00002	.88889	.11111	.0000	.00001	.00	11.50
137	0	8	.00001	1.00000	.00000	.0000	.00001	.00	11.88
138	0	8	.00001	1.00000	.00000	.0000	.00001	.00	10.88
139	0	8	.00001	1.00000	.00000	.0000	.00001	.00	9.88
140	0	8	.00001	1.00000	.00000	.0000	.00001	.00	9.88
141	1	8	.00001	.78500	.12500	.0000	.00001	.00	7.88
142	0	7	.00001	1.00000	.00000	.0000	.00001	.00	7.93
143	1	7	.00001	.85714	.14286	.0000	.00001	.00	6.93
144	0	6	.00001	1.00000	.00000	.0000	.00001	.00	7.00
145	2	6	.00001	.66667	.33333	.0000	.00001	.00	6.00
146	1	4	.00001	.75000	.25000	.0000	.00001	.00	7.75
147	0	3	.00001	1.00000	.00000	.0000	.00001	.00	9.17
148	1	3	.00001	.66667	.33333	.0000	.00000	.00	8.17
149	0	2	.00000	1.00000	.00000	.0000	.00000	.00	11.00
150	0	2	.00000	1.00000	.00000	.0000	.00000	.00	10.00
151	0	2	.00000	1.00000	.00000	.0000	.00000	.00	9.00
152	0	2	.00000	1.00000	.00000	.0000	.00000	.00	8.00
153	0	2	.00000	1.00000	.00000	.0000	.00000	.00	7.00
154	0	2	.00000	1.00000	.00000	.0000	.00000	.00	6.00
155	1	2	.00000	.50000	.50000	.0000	.00000	.00	5.00
156	0	1	.00000	1.00000	.00000	.0000	.00000	.00	8.50
157	0	1	.00000	1.00000	.00000	.0000	.00000	.00	7.50
158	0	1	.00000	1.00000	.00000	.0000	.00000	.00	6.50
159	0	1	.00000	1.00000	.00000	.0000	.00000	.00	5.50
160	0	1	.00000	1.00000	.00000	.0000	.00000	.00	4.50
161	0	1	.00000	1.00000	.00000	.0000	.00000	.00	3.50
162	0	1	.00000	1.00000	.00000	.0000	.00000	.00	2.50
163	0	1	.00000	1.00000	.00000	.0000	.00000	.00	1.50
164	1	1	.00000	.00000	1.00000	.0000	.00000	.00	.50

Appendix 3
Life Table for 605,528 Female Medflies

Table A3-1. Life Table for 605,528 Female Medflies Reared in a Total of 167 Cages at the Moscamed Mass Rearing Facility in Tapachula, Chiapas, Mexico. Data Were Gathered for the Project Titled "Oldest-Old Mortality in the Mediterranean Fruit Fly" Funded by the National Institute on Aging (Unpublished Data from J. R. Carey, P. Liedo, D. Orozco, and J. W. Vaupel)

Age Class	Number Dying	Number Alive	Living at Age x	Fraction Surviving from x to x+1	Fraction Dying from x to x+1	Dying in Interval x to x+1	Days Lived in Interval	Days Lived Beyond Age x	Expectation of Life
x	D_x	N_x	l_x	p_x	q_x	d_x	L_x	T_x	e_x
(1)	(2)	(3)	(4)	(5)	(6)	(7)	(8)	(9)	(10)
0	0	605528	1.00000	1.00000	.00000	.0000	1.00000	19.58	19.58
1	835	605528	1.00000	.99862	.00138	.0014	.99931	18.58	18.58
2	2347	604693	.99862	.99612	.00388	.0039	.99668	17.58	17.60
3	3132	602346	.99475	.99480	.00520	.0052	.99216	16.58	16.67
4	4218	599214	.98957	.99296	.00704	.0070	.98609	15.59	15.75
5	5042	594996	.98261	.99153	.00847	.0083	.97844	14.60	14.86
6	6310	589954	.97428	.98930	.01070	.0104	.96907	13.62	13.98
7	8460	583644	.96386	.98550	.01450	.0140	.95687	12.66	13.13
8	10712	575184	.94989	.98138	.01862	.0177	.94104	11.70	12.32
9	14862	564472	.93220	.97367	.02633	.0245	.91993	10.76	11.54
10	19897	549610	.90765	.96380	.03620	.0329	.89122	9.84	10.84
11	26697	529713	.87480	.94960	.05040	.0441	.85275	8.95	10.23
12	20252	503016	.83071	.93986	.06014	.0500	.80573	8.09	9.74
13	38152	472764	.78075	.91930	.08070	.0630	.74924	7.29	9.33
14	38671	434612	.71774	.91102	.08898	.0639	.68581	6.54	9.11
15	40015	395941	.65388	.89894	.10106	.0661	.62084	5.85	8.95
16	37015	355926	.58779	.89600	.10400	.0611	.55723	5.23	8.90
17	32405	318911	.52667	.89839	.10161	.0535	.49991	4.67	8.88
18	29307	286506	.47315	.89771	.10229	.0484	.44895	4.18	8.82
19	25789	257199	.42475	.89973	.10027	.0426	.40346	3.73	8.77
20	24186	231410	.38216	.89548	.10452	.0399	.36219	3.32	8.69
21	21579	207224	.34222	.89587	.10413	.0356	.32440	2.96	8.65
22	19352	185645	.30658	.89576	.10424	.0320	.29060	2.64	8.60
23	17942	166293	.27462	.89211	.10789	.0296	.25981	2.25	8.54
24	15840	148351	.24499	.89323	.10677	.0262	.23191	2.09	8.51
25	14085	132511	.21884	.89371	.10629	.0233	.20721	1.85	8.47
26	13930	118426	.19557	.88237	.11763	.0230	.18407	1.65	8.42
27	11874	104496	.17257	.88637	.11363	.0196	.16277	1.46	8.47
28	11334	92622	.15296	.87763	.12237	.0187	.14360	1.30	8.50
29	8647	81288	.13424	.89363	.10637	.0143	.12710	1.16	8.61
30	7342	72641	.11996	.89893	.10107	.0121	.11390	1.03	8.58

APPENDIX

Table A3-1. (*Continued*)

Age Class	Number Dying	Number Alive	Living at Age x	Fraction Surviving from x to x+1	Fraction Dying from x to x+1	Dying in Interval x to x+1	Days Lived in Interval	Days Lived Beyond Age x	Expectation of Life
x	D_x	N_x	l_x	p_x	q_x	d_x	L_x	T_x	e_x
(1)	(2)	(3)	(4)	(5)	(6)	(7)	(8)	(9)	(10)
31	7024	65299	.10784	.89243	.10757	.0116	.10204	.92	8.49
32	5807	58275	.09624	.90035	.09965	.0096	.09144	.81	8.45
33	5531	52468	.08665	.89458	.10542	.0091	.08208	.72	8.33
34	4973	46937	.07751	.89405	.10595	.0082	.07341	.64	8.25
35	4738	41964	.06930	.88709	.11291	.0078	.06539	.57	8.17
36	4135	37226	.06148	.88892	.11108	.0068	.05806	.50	8.14
37	3991	33091	.05465	.87939	.12061	.0066	.05135	.44	8.10
38	2958	29100	.04806	.89835	.10165	.0049	.04561	.39	8.14
39	2868	26142	.04317	.89029	.10971	.0047	.04080	.35	8.01
40	2493	23274	.03844	.89288	.10712	.0041	.03638	.30	7.93
41	2247	20781	.03432	.89187	.10813	.0037	.03246	.27	7.82
42	2012	18534	.03061	.89144	.10856	.0033	.02895	.24	7.71
43	1938	16522	.02729	.88270	.11730	.0032	.02569	.21	7.59
44	1895	14584	.02408	.87006	.12994	.0031	.02252	.18	7.53
45	1556	12689	.02096	.87737	.12263	.0026	.01967	.16	7.58
46	1233	11133	.01839	.88925	.11075	.0020	.01737	.14	7.57
47	1182	9900	.01635	.88061	.11939	.0020	.01537	.12	7.45
48	1079	8718	.01440	.87623	.12377	.0018	.01315	.11	7.39
49	874	7639	.01262	.88559	.11441	.0014	.01189	.09	7.37
50	935	6765	.01117	.86179	.13821	.0015	.01040	.08	7.26
51	704	5830	.00963	.87925	.12075	.0012	.00905	.07	7.34
52	682	5126	.00847	.86695	.13305	.0011	.00790	.06	7.28
53	636	4444	.00734	.85689	.14311	.0011	.00681	.05	7.32
54	466	3808	.00629	.87763	.12237	.0008	.00590	.05	7.46
55	472	3342	.00552	.85877	.14123	.0008	.00513	.04	7.43
56	417	3870	.00474	.85470	.14530	.0007	.00440	.04	7.57
57	329	2453	.00405	.86588	.13412	.0005	.00378	.03	7.77
58	321	2124	.00351	.84887	.15113	.0005	.00324	.03	7.89
59	269	1803	.00298	.85080	.14920	.0004	.00276	.02	8.21
60	192	1534	.00253	.87484	.12516	.0003	.00237	.02	8.56
61	175	1342	.00222	.86960	.13040	.0003	.00207	.02	8.72
62	136	1167	.00193	.88346	.11654	.0002	.00181	.02	8.95
63	137	1031	.00170	.86712	.13288	.0002	.00159	.02	9.06
64	137	894	.00148	.84676	.15324	.0002	.00136	.01	9.37
65	125	757	.00125	.83487	.16513	.0002	.00115	.01	9.98
66	79	632	.00104	.87500	.12500	.0001	.00098	.01	10.86
67	79	553	.00091	.85714	.14286	.0001	.00085	.01	11.34
68	51	474	.00078	.89241	.10759	.0001	.00074	.01	12.14
69	49	423	.00070	.88416	.11584	.0001	.00066	.01	12.54
70	32	374	.00062	.91444	.08556	.0001	.00059	.01	13.12
71	42	342	.00056	.87719	.12281	.0001	.00053	.01	13.30
72	31	300	.00050	.89667	.10333	.0001	.00047	.01	14.10
73	30	269	.00044	.88848	.11152	.0000	.00042	.01	14.66
74	25	239	.00039	.89540	.10460	.0000	.00037	.01	15.44
75	21	214	.00035	.90187	.09813	.0000	.00034	.01	16.19
76	20	193	.00032	.89637	.10363	.0000	.00030	.01	16.89
77	11	173	.00029	.93642	.06358	.0000	.00028	.01	17.79
78	15	162	.00027	.90741	.09259	.0000	.00026	.00	17.96

Table A3-1. (Continued)

Age Class	Number Dying	Number Alive	Living at Age x	Fraction Surviving from x to x+1	Fraction Dying from x to x+1	Dying in Interval x to x+1	Days Lived in Interval	Days Lived Beyond Age x	Expectation of Life
x	D_x	N_x	l_x	p_x	q_x	d_x	L_x	T_x	e_x
(1)	(2)	(3)	(4)	(5)	(6)	(7)	(8)	(9)	(10)
79	10	147	.00024	.93197	.06803	.0000	.00023	.00	18.74
80	8	137	.00023	.94161	.05839	.0000	.00022	.00	19.08
81	12	129	.00021	.90698	.09302	.0000	.00020	.00	19.23
82	9	127	.00019	.92308	.07692	.0000	.00019	.00	20.15
83	10	108	.00018	.90741	.09259	.0000	.00017	.00	20.79
84	7	98	.00016	.92857	.07143	.0000	.00016	.00	21.86
85	6	91	.00015	.93407	.06593	.0000	.00015	.00	22.50
86	3	85	.00014	.96471	.03529	.0000	.00014	.00	23.05
87	8	82	.00014	.90244	.09756	.0000	.00013	.00	22.88
88	6	74	.00012	.91892	.08108	.0000	.00012	.00	24.30
89	0	68	.00011	1.00000	.00000	.0000	.00011	.00	25.40
90	3	68	.00011	.95588	.04412	.0000	.00011	.00	24.00
91	1	65	.00011	.98462	.01538	.0000	.00011	.00	24.50
92	6	64	.00011	.90625	.09375	.0000	.00010	.00	23.88
93	4	58	.00010	.93103	.06897	.0000	.00009	.00	25.29
94	2	54	.00009	.96296	.03704	.0000	.00009	.00	26.13
95	3	52	.00009	.94231	.05769	.0000	.00008	.00	26.12
96	1	49	.00008	.97959	.02041	.0000	.00008	.00	26.68
97	1	48	.00008	.97917	.02083	.0000	.00008	.00	26.23
98	3	47	.00008	.93617	.06383	.0000	.00008	.00	25.78
99	0	44	.00007	1.00000	.00000	.0000	.00007	.00	26.50
100	0	44	.00007	1.00000	.00000	.0000	.00007	.00	25.50
101	0	44	.00007	1.00000	.00000	.0000	.00007	.00	24.50
102	4	44	.00007	.90909	.09091	.0000	.00007	.00	23.50
103	1	40	.00007	.97500	.02500	.0000	.00007	.00	24.80
104	2	39	.00006	.94872	.05128	.0000	.00006	.00	24.42
105	1	37	.00006	.97297	.02703	.0000	.00006	.00	24.72
106	1	36	.00006	.97222	.02778	.0000	.00006	.00	24.39
107	0	35	.00006	1.00000	.00000	.0000	.00006	.00	24.07
108	1	35	.00006	.97143	.02857	.0000	.00006	.00	23.07
109	1	34	.00006	.97059	.02941	.0000	.00006	.00	22.74
110	1	33	.00005	.96970	.03030	.0000	.00005	.00	22.41
111	1	32	.00005	.96875	.03125	.0000	.00005	.00	22.09
112	2	31	.00005	.93548	.06452	.0000	.00005	.00	21.79
113	1	29	.00005	.96552	.03448	.0000	.00005	.00	22.26
114	0	28	.00005	1.00000	.00000	.0000	.00005	.00	22.04
115	1	28	.00005	.96429	.03571	.0000	.00005	.00	21.04
116	0	27	.00004	1.00000	.00000	.0000	.00004	.00	20.80
117	0	27	.00004	1.00000	.00000	.0000	.00004	.00	19.80
118	0	27	.00004	1.00000	.00000	.0000	.00004	.00	18.80
119	1	27	.00004	.96296	.03704	.0000	.00004	.00	17.80
120	2	26	.00004	.92308	.07692	.0000	.00004	.00	17.46
121	1	24	.00004	.95833	.04167	.0000	.00004	.00	17.88
122	2	23	.00004	.91304	.08696	.0000	.00004	.00	17.63
123	0	21	.00003	1.00000	.00000	.0000	.00003	.00	18.26
124	2	21	.00003	.90476	.09524	.0000	.00003	.00	17.26
125	3	19	.00003	.84211	.15789	.0000	.00003	.00	18.03
126	1	16	.00003	.93750	.06250	.0000	.00003	.00	20.31

APPENDIX

Table A3-1. (*Continued*)

Age Class	Number Dying	Number Alive	Living at Age x	Fraction Surviving from x to x+1	Fraction Dying from x to x+1	Dying in Interval x to x+1	Days Lived in Interval	Days Lived Beyond Age x	Expectation of Life
x	D_x	N_x	l_x	p_x	q_x	d_x	L_x	T_x	e_x
(1)	(2)	(3)	(4)	(5)	(6)	(7)	(8)	(9)	(10)
127	1	15	.00002	.93333	.06667	.0000	.00002	.00	20.63
128	1	14	.00002	.92857	.07143	.0000	.00002	.00	21.07
129	2	13	.00002	.84615	.15385	.0000	.00002	.00	21.65
130	0	11	.00002	1.00000	.00000	.0000	.00002	.00	24.50
131	0	11	.00002	1.00000	.00000	.0000	.00002	.00	23.50
132	0	11	.00002	1.00000	.00000	.0000	.00002	.00	22.50
133	0	11	.00002	1.00000	.00000	.0000	.00002	.00	21.50
134	0	11	.00002	1.00000	.00000	.0000	.00002	.00	20.50
135	0	11	.00002	1.00000	.00000	.0000	.00002	.00	19.50
136	0	11	.00002	1.00000	.00000	.0000	.00002	.00	18.50
137	0	11	.00002	1.00000	.00000	.0000	.00002	.00	17.50
138	0	11	.00002	1.00000	.00000	.0000	.00002	.00	16.50
139	0	11	.00002	1.00000	.00000	.0000	.00002	.00	15.50
140	0	11	.00002	1.00000	.00000	.0000	.00002	.00	14.50
141	1	11	.00002	.90909	.09091	.0000	.00002	.00	13.50
142	1	10	.00002	.90000	.10000	.0000	.00002	.00	13.80
143	1	9	.00001	.88889	.11111	.0000	.00001	.00	14.28
144	0	8	.00001	1.00000	.00000	.0000	.00001	.00	15.00
145	0	8	.00001	1.00000	.00000	.0000	.00001	.00	14.00
156	0	8	.00001	1.00000	.00000	.0000	.00001	.00	13.00
147	1	8	.00001	.87500	.12500	.0000	.00001	.00	12.00
148	0	7	.00001	1.00000	.00000	.0000	.00001	.00	12.64
149	0	7	.00001	1.00000	.00000	.0000	.00001	.00	11.64
150	1	7	.00001	.85714	.14286	.0000	.00001	.00	10.64
151	0	6	.00001	1.00000	.00000	.0000	.00001	.00	11.33
152	0	6	.00001	1.00000	.00000	.0000	.00001	.00	10.33
153	1	6	.00001	.83333	.16667	.0000	.00001	.00	9.33
154	1	5	.00001	.80000	.20000	.0000	.00001	.00	10.10
155	0	4	.00001	1.00000	.00000	.0000	.00001	.00	11.50
156	0	4	.00001	1.00000	.00000	.0000	.00001	.00	10.50
157	0	4	.00001	1.00000	.00000	.0000	.00001	.00	9.50
158	1	4	.00001	.75000	.25000	.0000	.00001	.00	8.50
159	0	3	.00000	1.00000	.00000	.0000	.00000	.00	10.17
160	0	3	.00000	1.00000	.00000	.0000	.00000	.00	9.17
161	0	3	.00000	1.00000	.00000	.0000	.00000	.00	8.17
162	0	3	.00000	1.00000	.00000	.0000	.00000	.00	7.17
163	0	3	.00000	1.00000	.00000	.0000	.00000	.00	6.17
164	1	3	.00000	.66667	.33333	.0000	.00000	.00	5.17
165	0	2	.00000	1.00000	.00000	.0000	.00000	.00	6.50
166	0	2	.00000	1.00000	.00000	.0000	.00000	.00	5.50
167	0	2	.00000	1.00000	.00000	.0000	.00000	.00	4.50
168	0	2	.00000	1.00000	.00000	.0000	.00000	.00	3.50
169	0	2	.00000	1.00000	.00000	.0000	.00000	.00	2.50
170	0	2	.00000	1.00000	.00000	.0000	.00000	.00	1.50
171	2	2	.00000	.00000	1.00000	.0000	.00000	.00	.50

Index

Abbott's correction, 32, 150–52
Age, chronological, 5, 8
Age, first reproduction, 102
Age class, 5, 13, 45
Age pyramid, 8
Age structure, 82, 96, 134, 164, 176
Age structure, region-specific, 135
Age structure, transience, 99
Age-specific
  event, 70
  mortality, 12–22, 182–203
  rates, 6
  reproduction, 45
Aphids, 116–17
Aphis fabae, 116–17
Apis mellifera, 19, 127
Apple maggot, 23–24, 27, 30, 31
Arithmetic increase, 77

Balancing equation, 78
Bioassay, 152–53
Biosteres tryoni, 169–71
Birth
  intervals, 57, 60, 63
  origin, 119–20
  per capita, 78

Cause of death, 22, 26
Change, demographic, 4
Clutch
  gross rate, 67, 71
Cochliomyia hominovorax, 161
Cohort, 5
Cohort synthetic, 11
Colony, 124, 135
Community ecology, 4
Competing risk, 23
Convergence, 102–04
Crossovers, 186–88
Crude rate model, 78–79
Crude rates, 6

Culex pipiens, 74–75
Cumulative parity, 44, 65, 68
Curve fitting, 155

Daily parity, 44, 65–66
Death, per capita, 78
Death rate, 107, 130
Demographic metabolism, 89
Demographic rates, 5
Demographic stochasticity, 113
Demography, 3
Descendents, 136
Deterministic, 135
Development, 146
Development time, 103
Discard age, 162, 171
Distribution, 4
Dose, 152
Doubling time, 79, 98

Echo, 99
Ecology, 4
Effective kill rate, pesticide, 150
Elimination of cause, 29
Entropy, 40, 191–92
Environmental stochasticity, 113, 116
Environmental variation, 114
Ergodicity, 134
  strong, 134
  weak deterministic, 135
  weak stochastic, 135
Estimation, 140, 192–95
Event, 5
Exact age, 5, 45
Expectation
  of a random variable, 36
  of a sample proportion, 35
  of a sample sum, 35–36
Expectation of life, 51–52, 70, 126–27
Expected fecundity, 58
Extinct, 113

INDEX

Fecundity, 43, 101
  mean age, 51, 56
Female dominant, 106
Finite rate of increase, 87
Fission, colony, 127
Force of mortality, 147, 191–92
Future reproduction, 57–58

Generation time, 129, 130
Genotype frequency, 135
Geometric increase, 77–79
Geometric mean, 80
Geometric summation, 143
Gompertz, 159
Gonotrophic cycle, 148–49
Gross clutch mean age, 71
Growth rate, 134, 140
Growth rates sequence, 80

Harvest rate, 161, 164, 169
Hatch
  gross, 50, 52
  mean age, 51
Hierarchical, 130, 135
Honeybees, 19, 20, 124, 128, 135
Honeybee, Africanized, 127
Honeybee queen, 126–29
Honeybee workers, 127

In-migration, gross, 9
Indispensable mortality, 29
Individual, 5
Initial conditions, 100
Insect mass rearing, 161, 178
Insect-days, 142–45
Interchange, gross, 9
Intervals, 6
Intrinsic
  birth rate, 87–88, 103
  death, 103
  rate of increase, 84, 89, 104
  rate of increase: analytical approximations, 85
  rates, 7
  sex ratio, 112, 135
Intrinsic death rate, 87
Intrinsic rate of increase, 83
Irreplaceable mortality, 29

Key factor analysis, 32
Kin collateral, 136
Kinship, 136

LD_{50}, 152–153
Least squares, 157

Leslie matrix, 92, 94–95, 123, 134
Life, cohort table, 11
Life course, 5
Life, current table, 11
Life cycle, 131
Life expectancy sex ratio, 186
Life history scaling, 155
Life table, 11
  entropy, 39
  functions, 12, 183
  medfly, 196–203
  radix, 11
  statistics, 33
Life table, abridged, 19–20
Life table, cohort, 11
Life table, complete, 15
Life table, complete cohort, 13
Life table, current, 11
Life table, reproductive, 71
Linear equation, 157
Logistic equation, 158
Lotka, 81, 83, 85, 86, 92, 109, 110
Louse, human, 14, 40

Maternity, 44
Maternity net, 45, 103
Mean age
  death, 37
  reproduction, 91
Mean generation time, 91
Mediterranean fruit fly, 52–53, 61–68, 153–54, 162–70, 178, 196–203
Mexican fruit fly, 72
Migrants, 9
Migration, 119
Migration, net, 9
Migration stream, 9
Mite-days, 143
Mortality, 19, 28, 29, 101
  proportional differences, 189
  sex differences, 188
  sex-specific, 184
  smoothing, 184
  doubling time (MRDT), 185
  sex ratio, 186
Mortality period, 12
Mortality rate doubling time, 185
Mosquito, 74–75, 148–49
Multiple decrement, 11, 22–23, 31–32, 152
Multiregional demography, 118, 135

Natality, 43
  net, 46, 50, 52
  gross, 46, 50, 52, 150
Net clutch mean age, 71

Net interchange, 9
  net rate, 67, 71
Nulliparous, 74, 148

Oldest Old, 182–83, 191
Out-migration, gross, 9

Parasitism, 154–55
Parasitoid, 168–70, 179
Parity, 74, 148–49
Parity distribution, 136
Parity progression, 43, 71
Parity projection, 136
Parity reproductive, 64
Parous-days, 148
  past, 56
  remaining, 56
  gross (GRR), 86–87
  net, 87
Paternity, 44
Pearson type I, 161
Pediculus humanus, 14–15
Per capita, reproductive rates, 44
Population, 3–4
Population, stable, 87
Population
  biology, 4
  change (in size), 8
  change (in space), 9
  distribution, 7
  dynamics, 4, 77
  ecology, 4
  genetics, 4
  momentum, 104
  rates, 77
  replacement, 104
  size, 7
  structure, 8
Population studies, 4
Population traits, 5
Probability, 33–34
Probit analysis, 32, 152, 154
Production rate, 164, 170
Progenitors, 136
Projections, 80, 92, 95, 120–21
  age-by-region, 123–25

Radix, of life table 12, 172
Random environments, 113
  rate, 107, 130
Recruitment, 43
Renewal rate, 43
Reproduction, 43
Reproduction, net, 170

Reproductive heterogeneity, 57, 60, 64
Reproductive interval, 44
Reproductive value, 91–93
Restricted rates, 6

Schedule, gross, 44
Schedule, net, 44
Screwworm, 161
Sensitivity analysis, 38
Sex ratio, 8, 109, 111, 113, 162
Sex ratio, adult, 111–12
Sex ratio, primary, 8, 46, 169
Sex ratio, stable, 106
Sex ratio secondary, 8
Sex-specific, 110
Sex-specific life tables, 109
Single decrement life table, 11
Size, 4
Social insects, 124
Spider mite, 34, 47, 79, 84, 88, 93, 94–96,
    110, 141–42, 146, 164, 168
Stable age distribution, 88–90, 102
Stable age distribution (SAD), 134
Stable net maternity, 103
Stable population model, 81
Stable theory, 106
Stable stage distribution, 91
Stage
  duration, 145
  structure, 97, 102, 134, 179
Standard distance, 8
Stationary, 134
Stochastic
  demography, 113
  rate of growth, 115
  varying, 134
Stream, net, 9
Structure, 4
Subpopulations, 79
Superorganism, 124
Survival sex ratio, 186
Survivorship period, 12
Swarming, 129
  fraction, 129–30
  ratio, 129–30
  size, 129

Temperature, 154, 157
Termite, 127
Trapezoidal method, 142
Turnover, 9
Two-region population, 120
Two-sex model, 106, 135

Variance, 33, 35, 37